GeoPlanet: Earth and Planetary Sciences

For further volumes:
http://www.springer.com/series/8821

Jerzy Sąsiadek
Editor

Aerospace Robotics

Selected Papers from I Conference on Robotics in Aeronautics and Astronautics

Springer

Editor
Jerzy Sąsiadek
Carleton University
Ottawa
Canada

The GeoPlanet: Earth and Planetary Sciences Book Series is in part a continuation of Monographic Volumes of Publications of the Institute of Geophysics, Polish Academy of Sciences, the journal published since 1962 (http://pub.igf.edu.pl/index.php).

ISSN 2190-5193 ISSN 2190-5207 (electronic)
ISBN 978-3-642-34019-2 ISBN 978-3-642-34020-8 (eBook)
DOI 10.1007/978-3-642-34020-8
Springer Heidelberg New York Dordrecht London

Library of Congress Control Number: 2013932857

Printed on acid-free paper

Springer is part of Springer Science+Business Media (www.springer.com)

Series Editors

Managing Editor

Advisory Board

Kon-Kee Liu
Institute of Hydrological
and Oceanic Sciences
National Central University Jhongli
Jhongli, Taiwan

Teresa Madeyska
Research Centre in Warsaw
Institute of Geological Sciences
Warszawa, Poland

Stanisław Massel
Institute of Oceanology
Polish Academy of Sciences
Sopot, Polska

Antonio Meloni
Instituto Nazionale di Geofisica
Rome, Italy

Evangelos Papathanassiou
Hellenic Centre for Marine Research
Anavissos, Greece

Kaja Pietsch
AGH University of Science and
Technology
Kraków, Poland

Dušan Plašienka
Prírodovedecká fakulta, UK
Univerzita Komenského
Bratislava, Slovakia

Barbara Popielawska
Space Research Centre
Polish Academy of Sciences
Warszawa, Poland

Tilman Spohn
Deutsches Zentrum für Luft-
und Raumfahrt
in der Helmholtz Gemeinschaft
Institut für Planetenforschung
Berlin, Germany

Krzysztof Stasiewicz
Swedish Institute of Space Physics
Uppsala, Sweden

Roman Teisseyre
Earth's Interior Dynamics Lab
Institute of Geophysics
Polish Academy of Sciences
Warszawa, Poland

Jacek Tronczynski
Laboratory of Biogeochemistry
of Organic Contaminants
IFREMER DCN_BE
Nantes, France

Steve Wallis
School of the Built Environment
Heriot-Watt University
Riccarton, Edinburgh
Scotland, UK

Wacław M. Zuberek
Department of Applied Geology
University of Silesia
Sosnowiec, Poland

Preface

Space exploration and activities requires new methods and technologies and often involve autonomous and robotics operations. The beginning of space exploration involved several automatic missions to inner and outer planets, as well as to our Moon. The applications of robots in space make many operations not only possible but also safe.

The human participation in Space exploration is often a reflection of our interest and curiosity in Cosmos and our surrounding World. Long distances between planets make human participation difficult, involving high risk or simply impossible. Therefore, automatic and robotic missions are destined to play a leading, ever increasing role in Space exploration.

Robotics in Space could perform three different roles: automatic exploration missions, mobile robots or rovers and robotics manipulators.

Automatic exploration missions involve spacecraft probes. In some missions, where landing at the other planet is planned, the spacecraft probe is designed as a lander equipped with robotic arm. The lander is stationary and does not move but has capabilities to collect ground and rock samples with robotic arm.

Mobile robots or rovers are robotic vehicles designed to explore surfaces of planets. Rovers could be autonomous or controlled remotely from its Control Centre. Such robots navigate across a planet, stop in numerous places of interest to investigate, take images and collect samples.

Robotic manipulators in Space perform very important role of handling parts, materials or minerals. Very often the manipulators are used for assembly (e.g., Space Station assembly operation) or handling spacecraft in and out of orbit. Space manipulators allow astronauts to perform their mission in orbit by lifting and carrying them at the tip of manipulator. Space manipulators play important role in handling materials around the Space Station and help in supplying, resupplying and building new modules for Space Station.

Poland has been developing its space capacity and expertise since 1970. The long pursued goal of joining European Space Agency was achieved in 2012, thus opening Polish Space community the door to participating in technologically advanced endeavours of the Agency. Space robotics belongs to the core of ESA activities in the exploration program and paves the way to future human missions to Mars. Poland can contribute significantly to ESA unmanned exploration projects building on the long tradition of research in general robotics, recognised expertise of many Polish scientific groups, young generation of scientists, engineers, specialists, and finally, on interest of many technical universities in Poland in extending their field of research to space.

From this perspective, consolidation of robotics and space communities in Poland is a natural process and should be realised by participation in joint projects, broad international cooperation and in the first place, trough scientific meetings and conferences.

In recognition of important role of robotics in Space, a special, dedicated Workshop on 'Space Robotics' was held in the Research Space Centre PAN in Warsaw, June 3–5, 2011.

This book is addressed to broad robotics and space community and will be of interests to scientists and researchers working in those fields.

Warsaw, July 2012

Marek Banaszkiewicz
Jerzy Sąsiadek

Contents

Space Robotics and its Challenges

Jerzy Sąsiadek

Abstract Space robotics is a fascinating field that was developed for space explorations and space missions. Human participation in space exploration has its limits that are related to human physical endurance. Long duration of flights and missions together with hostile environment in space limits human involvement. Unmanned and autonomous missions coupled with mission automation became necessary for successful exploration. This chapter reviews types of robots used in space, their main design features and possible applications. Brief review of space manipulators and space robots is presented and discussed.

1 Introduction

The last quarter of 20th century brought about rapid development in Space exploration and related strong demand for new technologies that would enable and support such endeavours.

Space robotics is one of the most important technologies that enable human exploration of outer space. There are essentially three types of robots used in space: manipulators, mobile and flying/floating robots (Sasiadek 1992, 1994).

The manipulators are robotic arms that are used for space operation, assembly and servicing. There are several manipulators that were used in the past in space. The most familiar are Remote Manipulator System (RMS) known also as Canada Arm or Shuttle Arm, Space Station Remote Manipulator System (SSRMS) and Special Purpose Dextrous Manipulator (SPDM) also known as Dextre.

J. Sąsiadek (✉)
Carleton University, Ottawa, ON K1S 5B6, Canada
e-mail: jurek_sasiadek@carleton.ca

J. Sąsiadek (ed.), *Aerospace Robotics*, GeoPlanet: Earth and Planetary Sciences,
DOI: 10.1007/978-3-642-34020-8_1, © Springer-Verlag Berlin Heidelberg 2013

The second group are mobile robots such as rovers and autonomous ground vehicles. The third group are flying and floating robots known as unmanned aerial vehicles (UAV) (Sasiadek 2009) and spacecraft robots (telerobotic servicer).

Space robotic manipulators can perform repetitive and lengthy tasks with reduced risk and improved performance and require fewer infrastructures than manned systems. Also, there is no need for life support systems. There are many application examples, as:

- maintenance, repair, and assembly;
- spacecraft deployment and retrieval;
- extravehicular activity support;
- shuttle inspection.

The cost of placing one kilogram of payload on the Earth orbit is quite high and amounts to as much as 23–25 K dollars. An economic consideration requires robots to be lightweight systems. Space robots are made of lightweight materials that are flexible and are prone for changing shape, bending, shaking and vibrations. Space based robots have flexible links and flexible joints. Positioning of end-point of robot with flexible links and flexible joints is a formidable challenge. Control system algorithms are complex and difficult to execute in real-time. Also, they require powerful processors with large operating memory. Those processors and other instrumentation must be certified for use in space. This equipment may or may not be available for space applications.

2 Space Manipulators

Space manipulators suffer from flexible effects in links and joints. The main problem in robotic manipulators is the positioning of its end-point. Link and joint flexibilities may introduce bending and vibrations that would cause instability in control systems. Control strategies developed for flexible robots are very complex and difficult to execute in real-time. There are several robotic manipulators that were or are being used in space. Three of them, however, are of special importance due to their size and design complexity. They are Remote Manipulator System (RMS) used on board of Space Shuttle, Space Station Remote Manipulator System (SSRMS) and Special Purpose Dextrous Manipulator (SPDM).

2.1 Remote Manipulator System

RMS, called also Canadarm1 or Shuttle Arm, is 15.3 m long, its mass is 408 kg, diameter is 38 cm and payload capability is 29,500 kg. This arm was used for many years as a main tool for cargo deliveries in space, in particular for loading and unloading cargo from Shuttle cargo bay on the Earth orbit (Sasiadek et al. 2002).

Fig. 1 Remote Manipulator System—RMS (courtesy of www.csa.ca)

It was also used for other operation, e.g., astronauts space walks and/or satellite retrieval. An image of RMS taken in space is shown in Figs. 1 and 2.

2.2 Space Station Remote Manipulator System

SSRMS, known also as Canadarm2, is 17.6 m long, its mass is 1,800 kg and a diameter is 35 cm. Payload capability is 116,000 kg. It has seven degrees of freedom (DOF). This arm is part of International Space Station orbiting the Earth.

Fig. 2 RMS viewed from the shuttle cargo bay (courtesy of www.csa.ca)

Fig. 3 Space Station Remote Manipulator System—SSRMS (courtesy of www.csa.ca)

The SSRMS is shown in Fig. 3. SSRMS can be placed on the Mobile Base System (MBS) attached to the structure of Space Station. Mobile Base System (MBS) has 1 DOF and can work jointly with SSRMS.

2.3 Special Purpose Dextrous Manipulator

SPDM is also known as Dextre. It was launched on March 11, 2008. This is a two arm robot. The length is 2.5 m, its mass is 1,662 kg, and mass lifting capacity is 600 kg. It has 15 DOF (Degree of Freedom). The general view of the robot is shown in Fig. 4. This two arm robot can be placed on the SSRMS tip and used jointly with SSRMS and MBS. The robotic system has complex kinematics. Together the total system MBS/SSRMS/SPDM has at least 23 DOF. The simultaneous control of all degrees of freedom is difficult and requires advance control systems.

3 Space Mobile Robots and Rovers

Space mobile robots were used for exploration of Earth Moon and exploration of outer planets, in particular Mars. However, the first robot used for exploration in space was the Lunokhod 1 that was deployed on November 17, 1970 on the Moon as part of Luna 17 Mission. Mass of the vehicle was 756 kg. Other interesting Moon rover, Lunokhod 2 (shown in Fig. 5) was deployed during the Luna 21

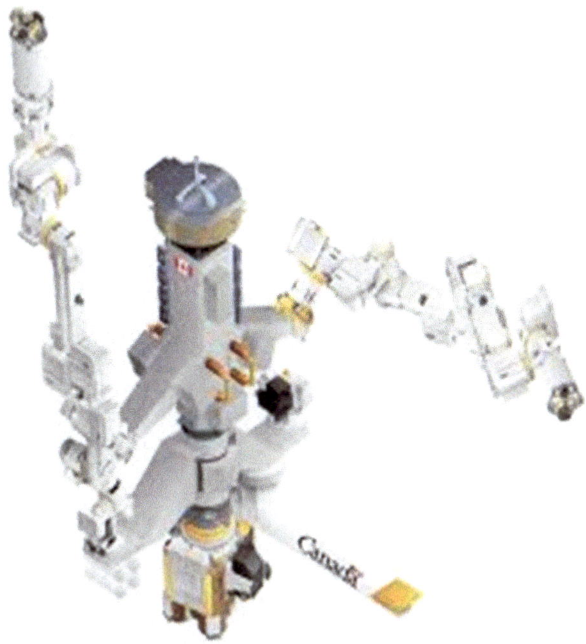

Fig. 4 Special Purpose Dexterous Manipulator—SPDM (courtesy of www.csa.ca)

mission on January 15, 1973. The mass of the rover was 840 kg and the mission itself lasted for around 130 days. Lunokhods Moon rovers were followed by succession of more or less advanced mobile robots that were used mostly for Mars exploration. The robots used in space were of several different types and designs. Some of them were landers, equipped with robotic arm that could not move from landing site but could perform several tasks like mineral finding and analysis. Viking Lander Missions with Orbiter 1 and Orbiter 2 stationary robots and Phoenix Lander were good examples of such robots. Viking 1 was launched on August 20, 1975 and landed on Mars on July 20, 1976 and Viking 2 was launched on September 9, 1975 and landed on Mars surface on September 3, 1976.

The Phoenix Mars Mission was launched in August 2007. The Phoenix Lander itself was a successor of Mars Surveyor 2001 developed by Lockheed-Martin.

More advanced robots were deployed on Mars on July 4, 1997. The Mars Pathfinder Mission was composed of the lander, Pathfinder, and rover called Sejourner (shown in Fig. 6).

The Mars Pathfinder Mission was launched on December 4, 1996 and landed on Mars on July 4, 1997. The Sejourner Mars Rover was a small and light vehicle. The rover was 65 cm long, 48 cm wide, and 30 cm tall. Its mass was 10.6 kg and maximum velocity was 1 cm/s or 10 mm/s. The maximum distance travelled was 100 m in total but never further than 16 m from the Pathfinder Lander. Sejourner was a small vehicle that was used for analyzing Mars rock composition and taking images along its path. The rover had its limitation due to the small size and limited

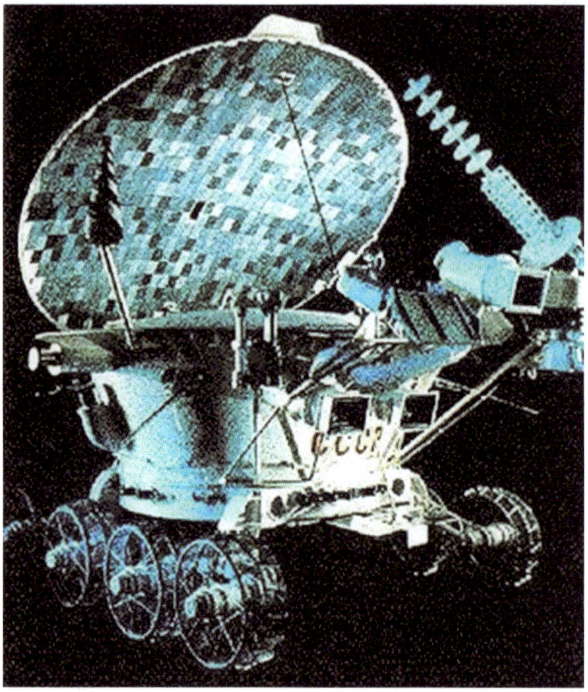

Fig. 5 Lunokhod 2 Moon Rover (courtesy of www.wikipedia.org)

Fig. 6 Sejourner Rover (courtesy of www.nasa.jpl.gov)

Fig. 7 The Spirit Mars Rover (courtesy of www.wikipedia.org)

Fig. 8 Curiosity Mars Rover (courtesy of www.wikipedia.org)

instrumentation and tools. It became obvious that larger and more powerful Mars rovers will be needed in future to perform more advanced exploration programs.

In January 4, 2004 the Spirit Mars Rover landed on Mars within Mars Exploration Rovers Mission. Three weeks later, another twin rover Opportunity arrived on the other side of Mars. The Spirit mission ended in 2010. Figure 7 shows the Spirit Mars Rover.

Spirit and Opportunity rovers were 1.6 m long, 2.3 m wide and 1.5 m high. Their mass was 180 kg. Both vehicles had six independently powered wheels. The maximum speed was 50 mm/s or 0.18 km/h. The average operational speed was 10 mm/s. The Spirit mission lasted 6 years and the rover travelled 7.73 km in total. Originally it was planned to cover 600 m distance. Control system was based on 20 MHz RAD 6000 CPU. At the end of its successful mission, after around 6 years of operation, the Spirit was stack in the sand between large rocks. It became clear that for more complex mission, even larger Mars rover vehicle will be needed.

On August 6, 2012, the new and most advanced Curiosity Mars Rover landed on the Mars. This rover is a six wheeled, car sized vehicle; its dimensions are: length 2.9 m, width 2.7 and height 2.2 m. The Curiosity mass was also significantly larger than in previous mission rovers. Its mass is 819 kg, with instrumentation 899 kg. The Curiosity robot is presented in Fig. 8.

The Curiosity Mars Rover mission is at present in its full operational mode.

References

Sasiadek JZ (1992) Space robotics and manipulators—lessons learned from the past and future missions and systems. In: Proceedings of the 12th IFAC international symposium on automatic control in aerospace control 92, Ottobrun, Germany, 7–11 Sept 1992 (invited plenary lecture)

Sasiadek JZ (1994) Space robotics and manipulators—the past and the future. IFAC Control Eng Pract J 2(4) Feb 1994

Sasiadek JZ (2009) Guidance, navigation, and control of autonomous vehicles. In: Proceedings of IFAC DECOM, Ohrid, Macedonia, Sept 2009 (invited plenary lecture)

Sasiadek JZ, Ollero R, Erbe H et al (2002) Robotics in next millenium. In: Proceedings of 16th IFAC world congress, July 2002, Elsevier, Barcelona, Spain

Comparing Locally Energy-Optimal Newton Algorithms to Steer Driftless Systems

Ignacy Duleba, Janusz Jakubiak and Michal Opalka

Abstract In this chapter two locally energy-optimal motion planning methods were compared. Both of them are based on the Newton algorithm (primarily exploited to solve an inverse kinematic task in robotic manipulators) adapted to motion planning of driftless nonholonomic systems. The first, continuous method employs a standard technique of optimization in the null-space of a Jacobian matrix. The matrix is derived from a linearization of the system along a current trajectory and the trajectory corresponds to a finite Fourier series representation of controls. The second, discrete one, admits only piecewise-constant controls modified at a finite number of sub-intervals of the time horizon. Some simulations performed on a model of the free-floating planar double pendulum placed atop of a base revealed that the continuous method is more reliable, although, when the goal configuration does not need to be reached very accurately, the discrete one can be useful and energy-effective. A classification of the discrete methods was also proposed. According to the classification the proposed discrete method is of the 1st order.

1 Introduction

Many models of contemporary robots (mobile platforms and robots, free-floating robots, under actuated manipulators), considered at a kinematic level, belong to a class of driftless nonholonomic systems described by the equations

$$\dot{\mathbf{q}} = \sum_{i=1}^{m} \mathbf{g_i}(\mathbf{q}) u_i \tag{1}$$

I. Duleba (✉) · J. Jakubiak · M. Opalka
Institute of Computer Engineering, Control and Robotics, Wroclaw University of Technology, Janiszewskiego Street 11/17 50-372 Wroclaw, Poland
e-mail: ignacy.duleba@pwr.wroc.pl

J. Sąsiadek (ed.), *Aerospace Robotics*, GeoPlanet: Earth and Planetary Sciences,
DOI: 10.1007/978-3-642-34020-8_2, © Springer-Verlag Berlin Heidelberg 2013

where $\mathbf{q} = (q_1, \ldots, q_n)^T$ is a configuration vector, $\mathbf{u} = (u_1, \ldots, u_m)^T$ are control inputs, and $\mathbf{g_1}, \mathbf{g_2}, \ldots, \mathbf{g_m}$ are smooth vector fields called generators of the system (1). A characteristic feature of the models (1) is that the number of state variables n is larger than the number of controls m. A motion planning task is to find controls $\mathbf{u}(t) = (u_1(t), \ldots, u_m(t))^T$ that steer the system (1) from a given initial configuration $\mathbf{q}_0$ to a final one $\mathbf{q}_f$. Despite the fact that the condition $m < n$ makes this task difficult, still there are many methods to solve it (Duleba 1998; LaValle 2006). The most effective were designed for special subclasses of driftless systems [flat (Rouchon et al. 1993), nilpotent (Lafferriere and Sussmann 1991), in a chain form (Murray and Sastry 1993)]. However, still there is a need for general-purpose methods, especially equipped with some additional features (optimality, collision avoidance). In this chapter one of the methods will be considered, namely the Newton algorithm. Originally, it was designed to solve inverse kinematic tasks in robot manipulators, then extended to driftless nonholonomic systems (Divelbiss and Wen 1993; Duleba and Sasiadek 2003) by appropriately redefined kinematics (Tchon and Jakubiak 2003). The main idea of the Newton algorithm applied to nonholonomic systems is quite simple: for given initial controls, define a trajectory initialized at $\mathbf{q}_0$, then linearize the system (1) along the trajectory, and, finally, modify the controls to force the end-point of the trajectory move towards the goal point $\mathbf{q}_f$. Applying this procedure iteratively, the goal configuration is reached if only the system (1) is controllable and a broad enough class of admissible controls is assumed. The main issue in modifying controls is their range. In continuous methods the controls are modified globally, over the whole time horizon, while for discrete methods more selective modifications of controls are possible. Both of the approaches have got some advantages and disadvantages. Continuous controls are usually easier to generate with power suppliers, while discrete methods are able to adjust trajectories only at those segments where it is the most necessary.

The chapter is organized as follows. In Sect. 2 two methods to determine modifications of controls will be discussed, both aimed at a local energy minimization (locality means that the energy optimal decisions are worked out for each iteration separately). The first (continuous) method modifies the whole trajectory by changing slightly coefficients of a selected basis of controls (for example, a Fourier basis but other continuous bases, like: Chebyshev, Legendre, Hermite, are also possible) from one iteration to another. The second (discrete) method admits only modifications of controls in a few selected sub-intervals (in this chapter only one among possible sub-intervals will be chosen to modify). In Sect. 3 simulations of the two methods carried out on the model of a planar double pendulum placed atop of a free-floating base are provided. Section 4 concludes the chapter.

2 Methods to Minimize Energy of Motion with the Use of the Newton Algorithm

Mathematically, the Newton algorithm is formalized as follows: for current controls $\mathbf{u}(\cdot)$ (initial controls are assumed) system (1) is linearized to get the form

$$\delta\dot{\mathbf{q}} = \mathbf{A}(t)\delta\mathbf{q} + \mathbf{B}(t)\delta\mathbf{u}, \tag{2}$$

with

$$\mathbf{A}(\mathrm{t}) = \frac{\partial(\mathbf{G}(\mathbf{q}(t))\mathbf{u}(t))}{\partial\mathbf{q}}, \quad \mathbf{B}(t) = \mathbf{G}(\mathbf{q}(t))$$

where $\delta\mathbf{w}$ introduces a small variation of quantity $\mathbf{w} \in \{\dot{\mathbf{q}}, \mathbf{q}, \mathbf{u}\}$ and $\mathbf{q}(\cdot)$ denotes a trajectory of the system (1) initialized at $\mathbf{q}_0$ generated with controls $\mathbf{u}(\cdot)$. The current end-point $\mathbf{q}(\mathrm{T})$ of the trajectory corresponding to controls $\mathbf{u}(\cdot)$ should be moved towards the goal $\mathbf{q}_f$. Therefore, controls $\mathbf{u}(\cdot)$ are modified by $\delta\mathbf{u}(\cdot)$ leading to the displacement of $\delta\mathbf{q}(\mathrm{T})$ given by

$$\delta\mathbf{q}(\mathrm{T}) = \int_0^T \mathbf{\Phi}(T,s)\mathbf{B}(s)\delta\mathbf{u}(\mathrm{s})ds, \quad (\delta\mathbf{q}(0) = 0), \tag{3}$$

where $\mathbf{\Phi}(t, s)$ is the fundamental matrix (Gantmacher 1988) satisfying the equation

$$\frac{d}{dt}\mathbf{\Phi}(t,s) = \mathbf{A}(t)\mathbf{\Phi}(t,s), \text{ with initial condition } \mathbf{\Phi}(s,s) = \mathbf{I}_n$$

($\mathbf{I}_n$ stands for the (n × n) identity matrix).

The kinematics of system (1) is the mapping

$$k_{\mathbf{q}_0,T} : L^2_m[0,T] \to R^n,\ k_{\mathbf{q}_0,T}(u(\cdot)) = \varphi_{\mathbf{q}_0,T}(u(\cdot)), \tag{4}$$

where $\mathbf{q}_0$ is an initial state, T is a fixed time horizon, and $L^2_m[0,T]$ is a Cartesian product of m (dim U = m) spaces of square integrable functions of time, defined on the interval [0, T]. The flow function (system trajectory) $\varphi_{\mathbf{q}_0,t}(u(\cdot))$ describes evolution of the state over the time period [0, t]. In this chapter, $\varphi_{\mathbf{q}_0,t}(u(\cdot))$ is also interpreted as the trajectory state reached at t.

In Eq. (3) one can influence $\delta\mathbf{u}(\cdot)$ to get desirable $\delta\mathbf{q}(T) = \xi(\mathbf{q}_f - \mathbf{q}(T))$, where a (small) positive value ξ is used to preserve validity of the linearization (2). This equation has got a nice physical interpretation, a final $\delta\mathbf{q}(T)$ can be viewed as a superposition of small displacements caused by $\delta\mathbf{u}(\mathrm{s})$, s ∈ [0, T] and transformed by the operator $\mathbf{\Phi}(\mathrm{T}, s)\mathbf{B}(s)$ from the time point s to the point T.

In the following subsections, two methods to determine $\delta\mathbf{u}(\cdot)$ will be reported.

2.1 Optimization of Energy in Null Space of the Jacobian Matrix

First method, well known in robotic literature (Duleba and Sasiadek 2003) is the following: a functional orthogonal basis is selected and controls are described in the form of a series

$$u_k(t) = \sum_{j=1}^{N^k} \lambda_j^k \phi_j^k(t), \quad k = 1, \ldots, m, \quad t \in [0, T], \tag{5}$$

where N^k is the number of items selected to express the kth control, $\phi^k(t)$ are some of the very first elements of the Fourier basis with its harmonics

$$(1, \cos(k\omega t), \sin(k\omega t))^T, \quad k = 1, \ldots, \quad \text{where } \omega = 2\pi/T.$$

In this way, the infinite dimensional functional space $L_m^2[0, T]$ was replaced by a finite dimensional space of coefficients λ, $\dim(\lambda) = \sum_{k=1}^{m} N^k$. It is worth noticing that different controls can be spanned by different harmonics (for example $u_1(t)$ can be a linear combination of the basis elements 1 and $\sin(\omega t)$ while $u_2(t)$ can be spanned by 1 and $\cos(2\omega t)$). The total energy of controls (5) is given by the formula

$$f(\lambda) = \frac{T}{2}\left(\sum_{k=1}^{m}\left\{2\left(\lambda_1^k\right)^2 + \sum_{i=2}^{N^k}\left(\lambda_j^k\right)^2\right\}\right), \tag{6}$$

where $\lambda^{\mathbf{k}} = (\lambda_1^k, \lambda_2^k, \ldots, \lambda_{N_k}^k)^T$, $k = 1, \ldots, m$ is the vector of coefficients of the kth control (5) and λ gathers all the vectors. Substituting (5) into (3) we get

$$\delta\mathbf{q}(T) = \int_0^T \mathbf{\Phi}(T, s)\mathbf{B}_{\boldsymbol{\lambda}}(s)ds \cdot \delta\lambda = \mathbf{J}_{q_0,T}(\mathbf{u}(\cdot, \lambda)) \cdot \delta\lambda, \tag{7}$$

where $\mathbf{J}_{q_0,T}(\mathbf{u}(\cdot, \boldsymbol{\lambda})) = \mathbf{J}_{q_0,T}(\boldsymbol{\lambda})$ denotes the Jacobian matrix and

$$\mathbf{B}_{\boldsymbol{\lambda}}(s) = \left[B_1\phi_1^1, \ldots, B_1\phi_{N_1}^1, B_2\phi_1^2, \ldots, B_2\phi_{N_2}^2, \ldots, B_m\phi_1^m, \ldots, B_m\phi_{N_m}^m\right],$$

where B_i, $i = 1, \ldots, m$ are columns of the matrix $\mathbf{B}$. The search of controls is performed within the space of their coefficients $\mathbf{\Lambda} \ni \lambda$ as the basis functions $\phi(t)$ are fixed. The vector λ is changed according to an iterative formula introducing the Newton algorithm for nonholonomic systems with the optimization in the null space of the Jacobian matrix (the last term in the following equation) (Nakamura 1991).

$$\lambda_{i+1} = \lambda_i + \xi_1 \cdot \mathbf{J}^{\#}_{\mathbf{q_0},\mathrm{T}}(\lambda_i)\left(\mathbf{q}_f - \mathbf{k}_{\mathbf{q_0},\mathbf{T}}(\mathbf{u}(\cdot,\lambda_i))\right)$$
$$- \xi_2\left(\mathbf{I} - \mathbf{J}^{\#}_{\mathbf{q_0},T}(\lambda_i)\mathbf{J}_{\mathbf{q_0},\mathrm{T}}(\lambda_i)\right)\frac{\partial f(\lambda)}{\partial \lambda}, \tag{8}$$

where $\mathbf{J}^{\#} = \mathbf{J}^{\mathrm{T}}(\mathbf{J}\mathbf{J}^{\mathrm{T}})^{-1}$ is the Moore–Penrose pseudo-inverse matrix inversion, $\mathbf{q}_f$ is a goal state, ξ_{i} are given coefficients and the initial value λ_0 is uniquely determined by the initial value of **u**.

The continuous version of the Newton algorithm for controllable systems (1) guarantees an existence, in each iteration, of such $\delta\mathbf{u}_{\mathrm{i}}$ that moves $\mathbf{q}_{\mathrm{i}}(T)$ towards $\mathbf{q}_f$. However, it changes controls in the whole interval [0, *T*]. It is possible that controls in some sub-intervals may disrupt effective approaching to the goal caused by the other pieces of controls. To avoid this drawback, a discrete way to modify controls will be proposed.

2.2 Optimization of Energy in One Sub-Interval of Controls

The other way to optimize energy of controls in one iteration progresses as follows: the interval [0, *T*] is partitioned into *N* sub-intervals and in each sub-interval only constant-value controls are admissible. In order to simplify implementation, the intervals are assumed to be of the same length, although from a theoretical point of view their length should be correlated with model-dependent quantities $\mathbf{A}(s)$, $\mathbf{B}(s)$, $\boldsymbol{\Phi}(T, s)$ on each sub-interval. In the s_ith sub-interval, $i \in \{1, \ldots, N\}$, a hyperplane spanned by column-vectors of the fixed $(n \times m)$ matrix $\boldsymbol{\Phi}(T, s_{\mathrm{i}})\mathbf{B}(s_{\mathrm{i}})\Delta s$, $\Delta s = T/N$, is defined. In this hyperplane one can move along by changing $\delta\mathbf{u}(s_{\mathrm{i}})$, aimed at getting desired $\delta\mathbf{q}(T)$. Unfortunately, $\delta\mathbf{q}(T)$ may not belong to the hyperplane as the hyperplane is *m*-dimensional and designed motion $\delta\mathbf{q}(T)$ lives in $n > m$ dimensional space. All what can be done is to project $\delta\mathbf{q}(T)$ into the hyper-plane, using the Gram-Schmidt orthogonalization procedure, to get decomposition $\delta q(s_i) = \delta q^{\perp}(s_{\mathrm{i}}) + \delta q^{\parallel}(s_{\mathrm{i}})$. The component $\delta q^{\parallel}(s_{\mathrm{i}})$ placed on the hyperplane can be compensated by admissible $\delta\mathbf{u}(s_{\mathrm{i}})$, while the $\delta\mathbf{q}^{\perp}(s_{\mathrm{i}})$ component cannot.

Those $\delta\mathbf{u}(s_{\mathrm{i}})$ when added to the current $\mathbf{u}(s_{\mathrm{i}})$ in the interval change the energy on motion (by increasing or decreasing it). This interval which causes the motion towards the goal (this requirement guarantees convergence of the process) with the maximal energy decrease is selected and used to change $\mathbf{u}(\cdot)$ for the next iteration. Formally, the criterion function which admits modifications of controls in a single sub-interval is given by the expression

$$K(s_i) = \left(\langle \mathbf{u}(s_i), \mathbf{u}(s_i)\rangle - \langle \mathbf{u}(s_i) + \delta\mathbf{u}(s_i), \mathbf{u}(s_i) + \mathbf{u}(s_i)\rangle\right) \cdot \left\| \frac{\delta\mathbf{q}^{\parallel}(s_i)}{\delta\mathbf{q}(s_i)} \right\|, K(s^*)$$
$$= \max_{i=1,\ldots N} K(s_i), \tag{9}$$

where $\langle\cdot,\cdot\rangle$ denotes inner product while $\|\cdot\|$ is the Euclidean metrics and $s^{\star}$ denotes the optimal interval (really modified in the current iteration). The function $K(s_i)$ in (9) is a product of two terms: the first one measures positive or negative changes of energy due to $\delta\mathbf{u}(s_i)$ acting on the sub-interval corresponding to s_i, while the second one, taking a value from the range [0, 1], evaluates how effective approaching to the goal is. In practical situations those sub-intervals with very small component $\| \delta q^{\|}(s_i) \|$ should be excluded from the maximization process in (9) not to slow down the convergence process. Obviously, the two components in Eq. (9) can be weighted to form other criterion functions.

3 Simulations

Continuous and discrete methods that locally minimize the energy of motion were compared on the model of a free-floating space robot depicted in Fig. 1 and characterized by the data $m_0 = 5$, $m_1 = m_2 = 1$, $l_1 = l_2 = 1$, $a = 1$, $b = 0.6$. Using principles of linear and angular momentum (Dubowsky and Papadopoulos 1993) laws (initial momenta were assumed to vanish) the following equations were derived

$$\dot{\mathbf{q}} = \begin{bmatrix} \dot{q}_1 \\ \dot{q}_2 \\ \dot{q}_3 \end{bmatrix} = \begin{bmatrix} 1 \\ 0 \\ A_1(q_1,q_2) \end{bmatrix} u_1 + \begin{bmatrix} 0 \\ 1 \\ A_2(q_1,q_2) \end{bmatrix} u_2 = g_1(\mathbf{q})u_1 + g_2(\mathbf{q})u_2 = G(\mathbf{q})\mathbf{u}, \tag{10}$$

where $n = 3$, $m = 2$ and

$$\begin{aligned} A_1(q_1,q_2) &= -(76 + 135c_{q_1} + 33c_{q_2} + 45c_{q_1q_2})/A, \\ A_2(q_1,q_2) &= -(23 + 16.5c_{q_1} + 4.5c_{q_1q_2})/A \\ A &= 105.2 + 27c_{q_1} + 33c_{q_2} + 9c_{q_1q_2}, \quad \mathbf{u} = (\dot{q}_1,\dot{q}_2)^T. \end{aligned}$$

It is worth noticing that the linear momentum conservation law generates holonomic constraints while the angular momentum conservation law produces

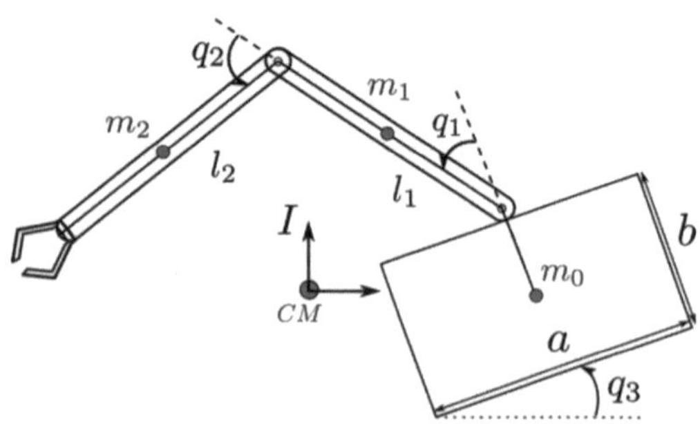

Fig. 1 Free-floating 2D pendulum placed atop of a base

nonholonomic constraints. Holonomic constraints were incorporated into the model to reduce the natural configuration space $(q_1, q_2, q_3, x, y)^T$ into the final one $(q_1, q_2, q_3)^T$. Based on the Chow theorem (Chow 1939) that provides the small-time local controllability condition $rank(LA(\mathbf{G}))(\mathbf{q}) = n$, it was verified that the model (10) satisfies the condition (vector fields $\mathbf{g}_1$ and $\mathbf{g}_2$ and their descendants (generated with the Lie brackets) span the configuration space everywhere) so it is also controllable.

Two motion planning tasks were considered with the following common data: the time horizon $T = 1$, the number of sub-intervals $N = 50$, initial controls $u_1(t) = u_2(t) = \cos(\omega t)$, $\omega = 2\pi/T$, $t \in [0, T]$ (for the discrete method appropriately replaced with piecewise-constant values). For the continuous method both controls were searched within the family $\lambda_1^k + \lambda_2^k \cos(\omega t) + \lambda_3^k \sin(\omega t), k = 1, 2,$ containing constant term and the first harmonics.

Task 1 was designed to move the robot from $\mathbf{q}_0 = (-45°, 90°, 60°)^T$ to $\mathbf{q}_f = (20°, 15°, 30°)^T$ while Task 2 from $\mathbf{q}_0 = (0°, 0°, 0°)^T$ to $q_f = (-90°, 60°, 45°)^T$. Resulting controls and trajectories for Task 1 were presented in Fig. 2 while for Task 2 in Fig. 3. In Fig. 4 a motion animation for Task 2 was provided. Numerical characteristics collected while solving the tasks were presented in Table 1.

From the simulations provided and those which were carried out but not reported here we can draw the following conclusions. The continuous method usually solves motion planning tasks with any accuracy assumed if only the number of parameters of controls is large enough comparing to the dimensionality of the configuration space (in our case the number of parameters was equal to $3 + 3 = 6$). However, for some tasks the energy expenditure on controls can be

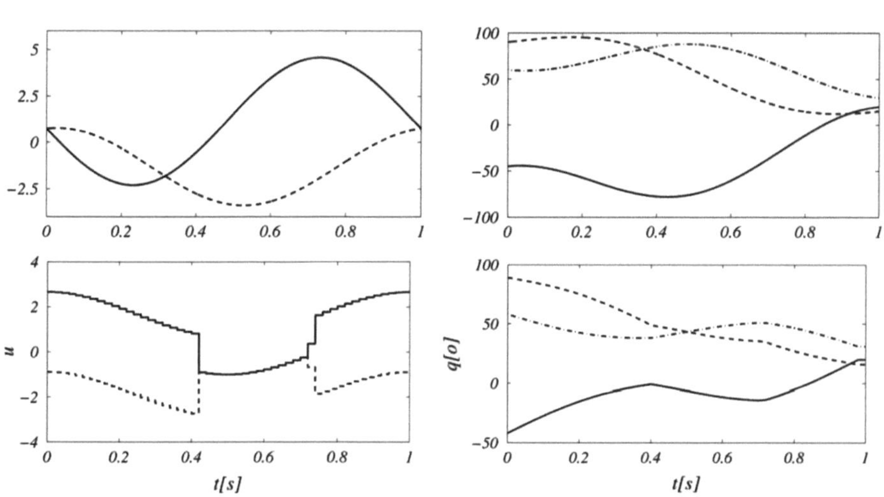

Fig. 2 Task 1: controls (*left column*, u_1/u_2 solid/slanted line) and trajectories (*right column*, $q_1/q_2/q_3$–solid/slanted/dotted line) generated with the continuous method (*first row*) and discrete one (*second row*)

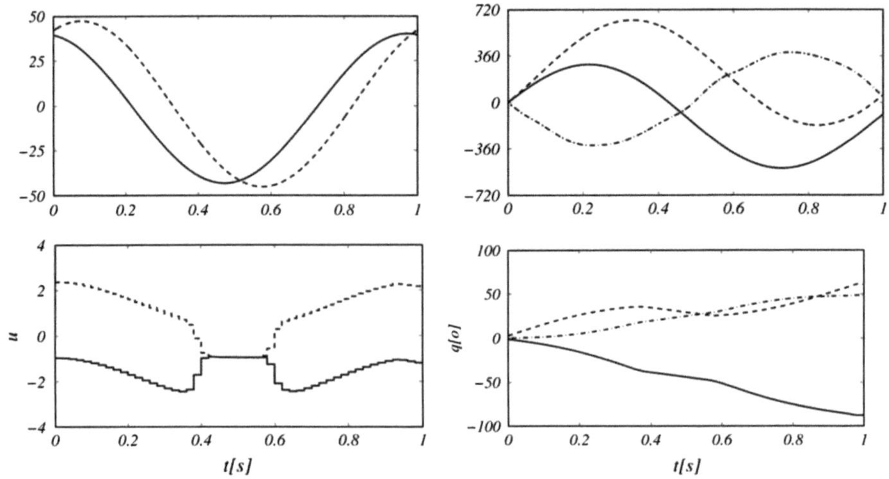

Fig. 3 Task 2: controls (*left column*, u_1/u_2 solid/slanted line) and trajectories (*right column*, $q_1/q_2/q_3$–solid/slanted/*dotted line*) generated with the continuous method (*first row*) and discrete one (*second row*)

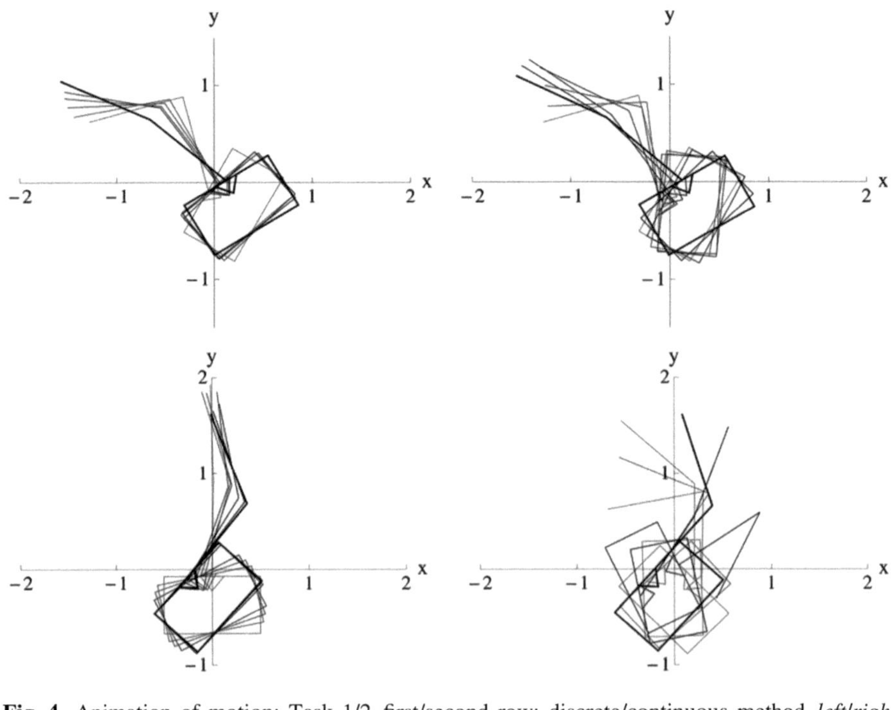

Fig. 4 Animation of motion: Task 1/2–first/second row; discrete/continuous method–*left*/*right* column

Table 1 Final error (err) in reaching the goal configuration, energy expenditure and the number of iteration to complete the computations, for Task 1 and Task 2 using the continuous and the discrete method

	Continuous			Discrete		
Task	Err(°)	Energy	Iter	Err(°)	Energy	Iter
1	8.9×10^{-9}	11.09	420	1.37	5.17	36
2	9.4×10^{-6}	1938.1	176	4.32	5.01	200

quite large (cf. Task 2). Also efficiency of solving the motion planning task can strongly depend on given initial controls.

The discrete method, admitting modification only one sub-interval of controls, does not allow to decrease the error to very small values. For some tasks (e.g., Task 1) the final error can be acceptable from a practical point of view but for the other ones (e.g., Task 2) not necessary. When the discrete method stops, all hyperplanes spanned by $\mathbf{\Phi}(T, s_i)\mathbf{B}(s_i)$ (for each s_i, $i = 1, \ldots, N$) are practically perpendicular to the $\delta\mathbf{q}(T)$, so controllable components $\delta q^{\|}(s_i)$ are very small. Consequently, no modifications performed on a single sub-interval can cause approaching the end-point of the trajectory towards the goal.

The discrete method can be improved by admitting multi-interval modifications but it can drastically increase computational complexity of the method. In the limit, when N tends to infinity, both of the methods (discrete and continuous) are the same. But also more subtle, and probably not so computationally expensive, modifications of the discrete method are possible. To explain this statement some Lie algebraic terms are indispensable.

Generators of the system (1) can be considered as formal Lie monomials and generators of a free Lie algebra (Serre 1964). With each formal generator its degree is associated. More complex Lie monomials are obtained using Lie bracket operator. A degree of a Lie monomial counts how many formal generators are used to get the monomial. Consequently, generators are degree one Lie monomials. Using the Campbell-Baker-Hausdorff-Dynkin formula (Duleba and Sasiadek 2001), higher degree Lie monomials (vector fields) can be generated by switching on and off piecewise-constant controls. For example, to generate the second degree vector field $[\mathbf{g}_1, \mathbf{g}_2]$ over the small time period Δs the following controls can be applied: $u_1 = (+1,0,-1,0)$, $u_2 = (0,+1,0,-1)$, and each interval lasts $\sqrt{\Delta s}$. From this example one can notice that motion along vector field $[\mathbf{g}_1, \mathbf{g}_2]$ is more energy consuming that a motion along generators $\mathbf{g}_1$, $\mathbf{g}_2$. A generalization of this observation for higher degree vector fields is also valid.

Now, the discrete method can be explained in Lie algebraic terms. The motions in each particular sub-interval characterized by s_i and coded by $\mathbf{B}(s_i)$ matrix can be viewed as a motion in linear spaces spanned by generators $\mathbf{g}_i$, $i = 1,\ldots, m$ of the system (1) and shifted by the operator $\mathbf{\Phi}(T, s_i)$. If they cannot help to solve the motion planning task, so, according to Chow's theorem (Chow 1939), they must be supported by higher degree vector fields $[\mathbf{g}_i, \mathbf{g}_j]$, $[\mathbf{g}_k, [\mathbf{g}_i, \mathbf{g}_j]]$, $i, j, k = 1,\ldots, m$. As the system (1) is controllable, so there exist some finite degree vector fields that force a motion of the end-point of a current trajectory towards the goal. With each discrete method its order can be associated. The order is inherited from a maximal degree vector field used to generate controls. According to this definition, the presented discrete method is of the first order. When more complex vector fields should be applied, an optimization procedure is used to determine at which time stamp s_i the generation of higher degree vector fields should be initialized. Scenarios of switching controls become even more cumbersome (for example, to generate third degree vector field 8 pieces of controls are used). Therefore, higher degree vector fields are to be generated only if lower degree vector fields fail to move $\mathbf{q}(T)$ towards $\mathbf{q}_f$.

Continuous and discrete methods should combine two contradictory factors: low computational complexity and a guaranteed convergence. By extending the number of parameters to change (number of harmonics in the continuous method or number of sub-intervals with varied controls) the chance of solving the task will increase but simultaneously computational complexity increases.

A singularity phenomenon is tightly bonded to an inverse kinematic task. For robot manipulators at singular configurations, the Jacobian matrix loses its maximal rank and the inverse task becomes ill-posed. For driftless nonholonomic systems (1) the problem of singularity is not so crucial. Firstly, because singular configurations cannot be computed analytically as the Jacobian matrix is based on the numerical data ($\boldsymbol{A}(s)$, $\boldsymbol{B}(s)$, $\boldsymbol{\Phi}(T, s)$). Secondly, the Jacobian matrix can be extended easily by adding some more harmonics to continuous controls (or applying higher order discrete methods). In typical case, extra columns of the Jacobian matrix allow to avoid singularities. However, the aforementioned methods to deal with singularities will increase the computational complexity. In practical situations a compromise should be made between flexibility in controls' representation and computational costs. However, the costs should not be overestimated as they are kinematic in nature (contrary to very extensive computations involving robot dynamics), usually low dimensional state spaces ($n \leq 6$) and some computations can be optimized (for example using the identity $\boldsymbol{\Phi}(T, s_1) = \boldsymbol{\Phi}(T, s)\cdot\boldsymbol{\Phi}(s, s_1)$, $s < s_1 < T$) and parallelized.

4 Conclusions

In this chapter two methods of nonholonomic motion planning for driftless systems based on the Newton algorithm were compared. The first, continuous one uses the Fourier basis to express controls and utilizes optimization in the null-space of the Jacobian matrix. When the representation of controls is rich enough, it solves motion planning tasks for controllable systems with any selected accuracy. For some tasks, to preserve the convergence of the algorithm, quite energy-expensive trajectories were generated with this method. For a discrete method, admitting modifications of controls only in one sub-interval, it may happen that the motion planning task cannot be solved with any assumed accuracy. The discrete method prefers energy-cheap trajectories. Some hints to improve convergence of the discrete method were also formulated. Both of the methods are local and strongly depend on initial controls assumed.

References

Chow WL (1939) Über systeme von linearen partiellen Differentialgleichungen erster Ordnung. Math Annalen 117(1):98–105

Divelbiss AW, Wen J (1993) Nonholonomic path planning with inequality constraints. IEEE Conf Decis Control 3:2712–2717

Dubowsky S, Papadopoulos E (1993) The kinematics, dynamics, and control of free-flying and free-floating space robotic system. IEEE Trans Rob Autom 9(5):531–542

Duleba I (1998) Algorithms of motion planning for nonholonomic robots. Publishing House of Wrocław University of Technology, Wroclaw

Duleba I, Sasiadek J (2001) Calibration of control in steering nonholonomic systems. Control Eng Pract 9(2):217–225

Duleba I, Sasiadek J (2003) Nonholonomic motion planning based on Newton algorithm with energy optimization. IEEE Trans Control Sys Technol 11(3):355–363

Gantmacher FR (1988) Matrix theory. Nauka, Moscow

Rouchon P, Fliess M, Levine J, Martin P (1993) Flatness, motion planning and trailer systems. IEEE Conf Decis Control 3:2700–2705

Lafferriere G, Sussmann H (1991) Motion planning for controllable systems without drift. IEEE Conf Robot Autom 2:1148–1153

LaValle SM (2006) Planning algorithms. Cambridge University Press, Cambridge

Murray RM, Sastry SS (1993) Nonholonomic motion planning: steering using sinusoids. IEEE Trans Autom Control 38(5):700–716

Nakamura Y (1991) Advanced robotics: redundancy and optimization. Addison Wesley, New York

Serre J-P (1964) Lie algebras and lie groups. Lecture notes in Mathematics, Springer

Tchon K, Jakubiak J (2003) Endogenous configuration space approach to mobile manipulators: a derivation and performance assessment of Jacobian inverse kinematics algorithms. Int J Control 26(14):1387–1419

Simulation and Visualization of Attitude Control System Operations for BRITE-PL Satellite

Grzegorz Juchnikowski, Jakub Lisowski, Konrad Skup, Roman Wawrzaszek, Tomasz Barciński and Tomasz Zawistowski

Abstract This chapter addresses the problem of satellite simulation software development. Such a tool was created as a part of BRITE-PL project, whose purpose is to prepare, launch and operate the first Polish scientific satellite. It allows engineers to virtually test different Attitude Determination and Control Systems (ADCS) and analyze satellite on-orbit behavior. Onboard sensors and actuators have been modeled and included in the software. Object oriented programming approach has been used, which makes the platform more flexible and simplifies further development.

1 Introduction

Nanosatellites gain a growing acceptance within space exploration community as efficient tools in performing a wide range of technical and scientific tasks.

They could be used as star-gazing telescopes, like BRITE constellation spacecraft (Moffat et al. 2006), or play an important role as damage-control verifiers for space vehicles like the volleyball-sized Miniature Autonomous Extravehicular Robotic Camera (Mini AERCam) (Fredrickson 2001). According to NASA, the Mini AERCam could provide beneficial on-orbit views that cannot be obtained from fixed cameras, cameras on robotic manipulators, or cameras carried

G. Juchnikowski · J. Lisowski (✉) · K. Skup · R. Wawrzaszek · T. Zawistowski
Space Research Centre of the Polish Academy of Sciences,
Bartycka 18A 00-716 Warsaw, Poland
e-mail: jilisowski@cbk.waw.pl

G. Juchnikowski
e-mail: jlisowski@cbk.waw.pl

J. Lisowski · T. Barciński
Department of Control and Measurements, West Pomeranian University of Technology,
Piastów 17 70-310 Szczecin, Poland

J. Sąsiadek (ed.), *Aerospace Robotics*, GeoPlanet: Earth and Planetary Sciences,
DOI: 10.1007/978-3-642-34020-8_3, © Springer-Verlag Berlin Heidelberg 2013

by space-walking crewmembers. For Shuttle or International Space Station missions, Mini AERCam could support external robotic operations by supplying situational awareness views to operators, supplying views of spacewalk operations to flight and/or ground crews, and carrying out independent visual inspections (Trinidad and Humphries 2005). Such sophisticated robotic implementation of nanosatellites is possible due to complex Attitude Control System, which can be additionally enhanced by orbit correction capabilities, enabling fully controlled maneuvering in space. In spite of small size, the accuracy of nanosatellite three-axis attitude control system is sufficient for conducting precise astronomical observations. Brite spacecraft ACS it can enable arcminute pointing performance.

Fundamental part of the satellite simulation tool is the rigid body dynamics and its associated equations of motion, which can be derived using elementary Newton laws of particle dynamics. Satellite system dynamics is addressed in Sect. 3. The problem is well known and has been described in many engineering and mathematical textbooks (Schaub and Junkins 2003; Angeles 1997; Murray et al. 1994). Specific application of rigid body equations in computer language has been proposed and well described by Baraff (2001), who also addresses the problem of constrained dynamics and rigid body contacts.

The market offers many commercial software tools for numerical modeling of rigid body or multi body behavior. Some of them are designed to perform very fast, not necessarily accurate, computations for real-time simulations used extensively in the computer game industry. Other tools target scientific applications, where computational accuracy is crucial and simulation time drops to second place. Available software includes the *Mathworks SimMechanics* toolbox for *MATLAB* product or *Bullet Physics Library*, which is an open source physics engine developed by Erwin Coumans et al. The question arises as to whether it is justified to create new software from scratch if many similar products are available.

Our effort is not only dedicated to create the simulation environment, but also to gain deeper insight into the laws of physics, which apply to satellite motion, software development process gives a great opportunity to do so. List of other important benefits includes the ability to design flexible platform which fits perfectly our needs.

First simulations with use of the tool described were performed for the satellite BRITE-PL.

2 Description of BRITE-PL

BRITE-PL is an ongoing project, whose purpose is to prepare, launch and operate the first Polish scientific satellite, designed to observe the brightest stars in our galaxy. Two satellites called *LEM* and *Heweliusz* will be the Polish contribution to the *BRITE* consortium. *BRITE* mission objectives are described in (Moffat et al. 2006). Aforementioned satellites are considered nanosatellites—the class of spacecrafts with weight between 1 and 10 kg and size 10 cm–1 m. Most of the subsystems for both satellites are produced by Space Flight Laboratory and

Sinclair Interplanetary, both from Toronto. Limited weight and size restrict the set of sensors and actuators, which together with computation unit are responsible for the spacecraft attitude control.

In the case of *BRITE-PL* three reaction wheels and three magnetic coils are used as actuators. Set of attitude sensors consists of six sun sensors, a three-axis magnetometer and a star tracker. The satellite lacks engines or any other linear momentum exchange devices, which would allow to affect the total momentum of the spacecraft. As a consequence, the satellite orbit is uncontrollable and depends on the initial conditions at the time of separation from launch vehicle. Main satellite subsystems are shown in Fig. 1. More technical information about *BRITE-PL* satellite can be found in (Orleański et al. 2010).

3 System Dynamics

This section is divided into two parts: the subsection describing the properties of rigid body motion dynamics and methods used to represent its attitude in space, and the part that is about the disturbances affecting orbiting spacecraft motion.

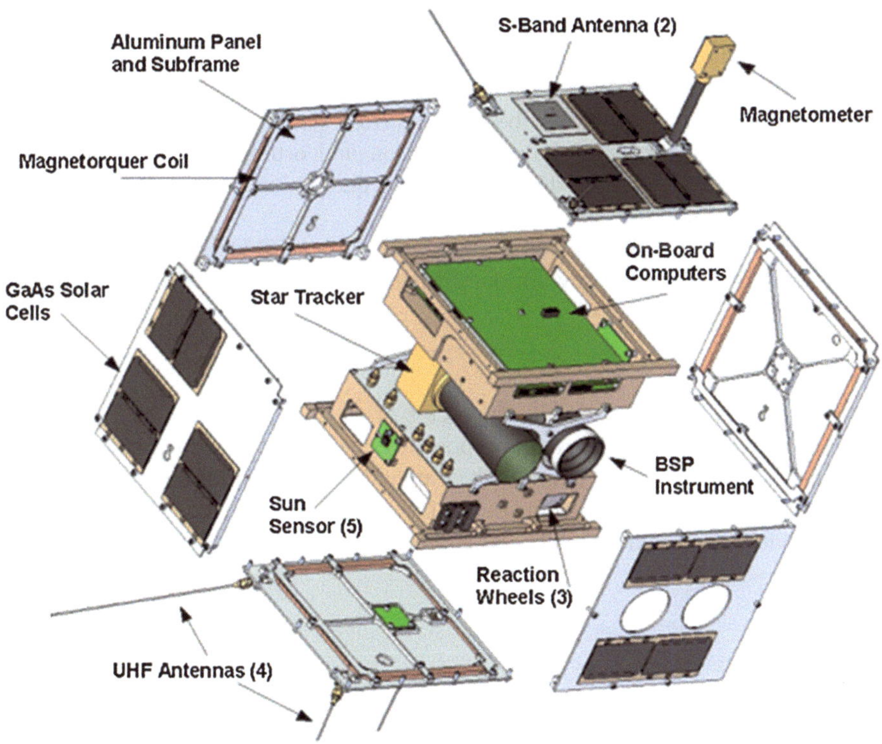

Fig. 1 Exploded view of BRITE-PL satellite. © UTIAS/SFL

3.1 Rigid Body Motion

From the mathematical point of view, rigid body motion is the transformation that preserves the distance between points and angles between vectors. It can be shown that such transformations are represented by the elements of Special Euclidean Group SE(3) (for the proof consult Murray et al. 1994 or Angeles 1997). These elements give relative displacement between frames and consist of two parts: translational part described by vector $\mathbf{p} \in \mathbb{R}^3$ and rotational part $\mathbf{R} \in SO(3)$.

In our case the control system is unable to generate the linear force acting on the satellite center of mass, so the differential equations describing its trajectory are omitted in model. However, the circular orbit model has been adopted to describe relative two body motion (Earth-satellite and Sun-Earth), which is necessary to estimate the Sun position in satellite fixed frame.

As has been mentioned before, rotational part of motion is described by an element $\mathbf{R} \in SO(3)$, which can be represented by 3×3 real matrix, if the orthonormal base has been defined. Instead of using nine redundant elements of rotation matrix to describe relative displacement, the matrix invariants can be used. Probably the most frequently used matrix invariants are the *Euler-Rodriguez parameters*, which have a compact representation using unit quaternions. Rotation matrix parametrization with Euler-Rodriguez parameters has been described in many different textbooks (Schaub and Junkins 2003; Angeles 1997). The comprehensive study of quaternion numbers and associated algebra can be found in (Vicci 2001; Pervin and Webb 1983 and Eberly 2010). Our simulation tool uses quaternion approach, because of better numerical stability. Rotation matrix may not be orthogonal after subsequent integration and orthogonalization is a computationally expensive task to do. In the case of quaternions a much less expensive normalization is needed.

Let $q \in \mathbb{H}$ be the unit quaternion describing the attitude of rigid body with respect to inertial frame. Its evolution $q(t)$ is described by the differential equation

$$\dot{q} = \frac{1}{2} q \omega, \tag{1}$$

where $\omega \in \mathbb{H}$ is the angular velocity vector of rigid body written as a pure quaternion (quaternion with the real component equals 0). All components of quantities in Eq. (1) are written in the body frame. The model of spacecraft dynamics with arbitrary number of reaction wheels has been derived using similar approach as in (Schaub and Junkins 2003) and can be written as

$$\begin{bmatrix} \mathbf{I} & i_a\mathbf{e}_1 & i_a\mathbf{e}_2 & \cdots \\ i_a\mathbf{e}_1^T & i_a & 0 & \cdots \\ i_a\mathbf{e}_2^T & 0 & i_a & \cdots \\ \vdots & \vdots & \vdots & \ddots \end{bmatrix} \begin{bmatrix} \dot{\omega} \\ \dot{\theta}_1 \\ \dot{\theta}_2 \\ \vdots \end{bmatrix} = \begin{bmatrix} -(\omega \times \mathbf{I}\omega) - \sum_i i_a\theta_i(\omega \times \mathbf{e}_i) \\ 0 \\ 0 \\ \vdots \end{bmatrix} + \begin{bmatrix} 0 \\ \tau_1 \\ \tau_2 \\ \vdots \end{bmatrix} + \begin{bmatrix} -(b_E \times m_c) \\ 0 \\ 0 \\ \vdots \end{bmatrix} + \begin{bmatrix} t_{ext} \\ 0 \\ 0 \\ \vdots \end{bmatrix} \tag{2}$$

where

$$\mathbf{I} = \mathbf{I}_s + \sum_i i_a\mathbf{e}_i\mathbf{e}_i^T + i_p\left(\mathbf{I}_3 - \mathbf{e}_i\mathbf{e}_i^T\right) \tag{3}$$

In the above equations, the following notation is used:

- $\boldsymbol{\omega}$ denotes satellite angular velocity vector.
- θ_i denotes angular speed of i-th reaction wheel with respect to satellite.
- τ_i denotes torque generated by i-th reaction wheel's motor.
- $\boldsymbol{m}_c$ denotes magnetic dipole vector generated by control coils.
- $\boldsymbol{t}_{ext}$ denotes external torque vector exerted on satellite body.
- $\mathbf{e}_i$ denotes vector indicating the axis of rotation of the i-th reaction wheel.
- i_a and i_p are respectively axial and perpendicular components of reaction wheel moment of inertia.
- $\mathbf{I}_s$ denotes satellite inertia tensor matrix.
- $\mathbf{b}_E$ denotes Earth magnetic field vector.
- $\mathbf{I}_3$ denotes 3×3 identity matrix.

All vector and tensor quantities are written in satellite's coordinate system. Eqs. (1) and (2) describe spacecraft's dynamical behavior and can be easily integrated with proper numerical scheme.

3.2 Disturbance Modeling

The torque $\boldsymbol{t}_{ext}$ from (2) is a composition of different disturbing torques

$$\boldsymbol{t}_{ext} = \boldsymbol{t}_p + \boldsymbol{t}_{gg} + \boldsymbol{t}_{pm} \tag{4}$$

Four kinds of disturbances are considered: aerodynamic drag, solar radiation pressure, gradient of gravity field, and parasitic magnetic dipole. The two first are treated the same way, since their nature is similar. Such disturbance torques are modeled as

$$\boldsymbol{t}_p = p(\mathbf{u} \times \mathbf{k}), \tag{5}$$

where $\boldsymbol{t}_{\mathrm{p}}$ corresponds to the torque being a result of an aerodynamic drag or solar radiation pressure, the scalar value p is a pressure, the unit vector $\mathbf{k}$ represents the direction of applied force (it opposes sun direction vector in case of solar radiation pressure and satellite velocity vector in case of aerodynamic drag) and the vector parameter $\mathbf{u}$ describes the net unbalance of spacecraft surface distribution with respect to its center of gravity.

According to (Hughes 2004) the gravity gradient torque can be calculated using the equation

$$\boldsymbol{t}_{gg} = 3\omega_o^2 \mathbf{R}^T \begin{bmatrix} -I_{23}^o & I_{13}^o & 0 \end{bmatrix}^T, \tag{6}$$

where parameter ω_{o} is angular rate of the orbit, I_{23}^o and I_{13}^o are inertia tensor components written in the orbit frame and $\mathbf{R} \in SO(3)$ describes spacecraft's attitude with respect to the inertial frame. It should be noted that $\mathbf{R}$ is obtained from spacecraft's attitude quaternion q (map from unit quaternion group to Special Orthogonal group is surjective).

Torque coming from parasitic magnetic dipole $\mathbf{m}_p$ in the Earth's field $\mathbf{b}_{\mathrm{E}}$ is equal to

$$\boldsymbol{t}_{pm} = \left(\mathbf{m}_p \times \mathbf{b}_E\right), \tag{7}$$

A magnitude of the parasitic dipole can be estimated by taking into account the geometry of harness subsystem and maximum wire currents. Ferromagnetic elements can also influence the net dipole. Not only the value of the dipole, but the way it changes in time is important. Simple simulation of the dipole as a Gaussian noise is far from reality and makes no sense. Time constants characteristic for the control algorithms used for attitude stabilization are rather long, and any noisy component of a disturbance will be averaged almost to zero in the time comparable to the time constant. On the other hand, considering parasitic dipole constant in time is also unrealistic. Kalman filter used in ADCS can estimate the value of a constant parasitic dipole and it can be compensated with magnetorquers. The following model, that implements the worst case of temporal changes, has been adopted to model the parasitic dipole

$$m_{p_i} = m_0 \sin \varphi_i\,, \quad i = 1, 2, 3 \tag{8}$$

where phases φ_i are results of integration $\varphi_i = \int\limits_0^t \omega_i(\tau)d\tau$ and frequencies ω_i are random numbers with average about $10^{-3}\mathrm{s}^{-1}$ and with spread about half of the average.

4 Implementation

The implementation of the simulator, under the code name *Busola*, is done using *MATLAB* environment. Control algorithms are implemented in C and compiled to dll libraries. This separation has a clear goal: this way the C code of ADCS flight software can be easily tested. These libraries are loaded by the *Busola* using internal *MATLAB's* interface to dll. The following subsections describe in more details the most important parts, components and ideas of the simulator. The simulator's *GUI* is shown on Fig. 2.

4.1 Object Oriented Programming

A typical satellite consists of several components, like Sun sensors, Star Tracker, Reaction Wheels etc. All these components work totally independent of each other. This means they can be set in different combinations depending on the satellite's purpose and mission objectives. Keeping in mind the last property we can get to the conclusion that the object oriented programming is ideal to be used in a universal simulator.

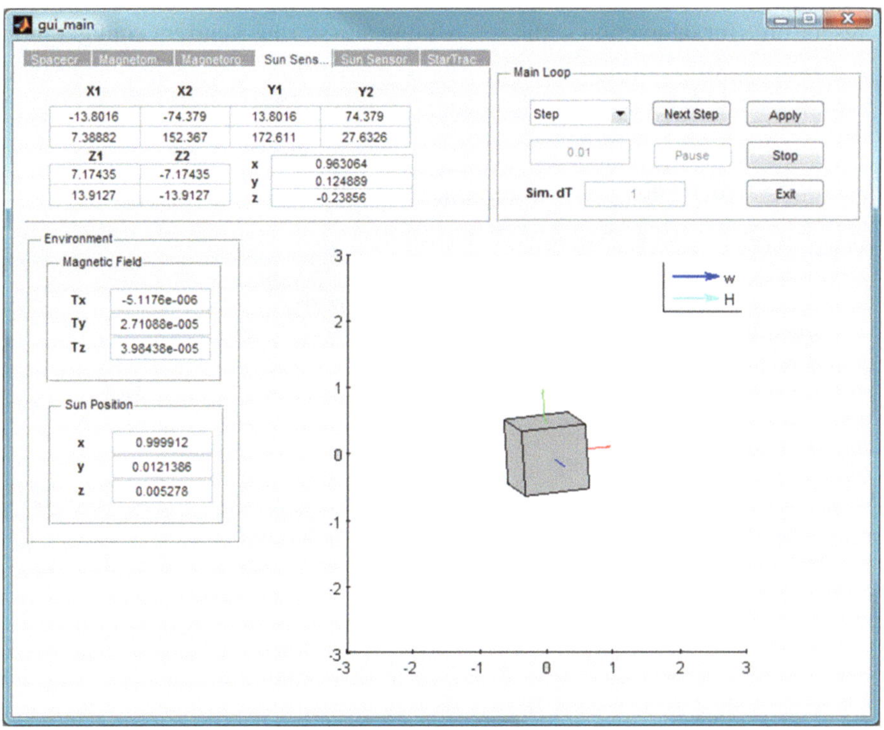

Fig. 2 Software's GUI

Implementing each of the satellite's components as a single and independent class gives the possibility of creating the numerical models of satellites already built as well as the configurations of satellites planned in the future. The former case can be used to verify the simulator itself. The latter makes the simulator an outstanding tool in aiding the process of controller design.

4.2 MATLAB Implementation

The decision to implement the *Busola* simulator in *MATLAB* is based on four key factors. Namely, *MATLAB*:

- has huge number of functions, including matrix operations, quaternion and vector operation support,
- supports object oriented programming,
- allows combined use of m files, dll libraries and *Simulink* models,
- allows to generate executable files from its source code.

The logical implementation of the *Busola* simulator is shown in Fig. 3. The simulator consists of four main groups of elements:

- *Busola*—It is the core of the simulator. It calls methods from other classes. It is responsible for interaction with user, reading and storing a current configuration.

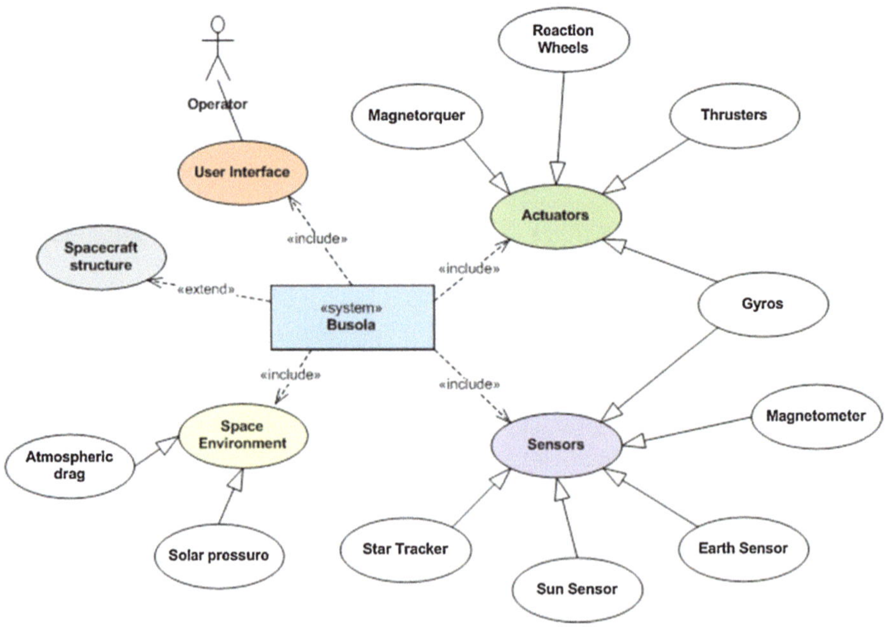

Fig. 3 Logical representation of the Busola simulator

It logs the simulation result to log files. The other important task this class is responsible for is numerical integration. The fixed step ODE4 *Runge–Kutta* scheme is used, because of accuracy needs fulfilment and implementation simplicity. This method does not guarantee motion invariants to be satisfied, but under the assumption that developed control system is stable invariants error is negligible.
- Actuators—It overwhelms all components that can generate torque and moments on the satellite. Applied torques on spacecraft's body are calculated with respect to actuators constraints and current system state.
- Sensors—This is a group of components that provide information from measurements to the simulator's internal control algorithms. An example how the sensor's work is implemented is shown on Fig. 6. Namely, the sensor gets actual measurement (in example the actual magnetic field). Vector components of measurement are transformed to the sensor's frame using spacecraft's attitude matrix and the information about relative sensor location in satellite body. Afterwards it is quantized and the noise is added. Finally, the output of the sensor is set and as a digital value it is forwarded to the simulator (Fig. 4).
- Environment—It is a group of external, environmental elements that acts on the satellite moving around the orbit.

4.3 C dll Algorithms

The satellite onboard systems are implemented in C programming language. In order to test the exact algorithms which will reach outer space with satellite flight model, the proper functionality has been added. Spacecraft software is compiled as *Windows* dynamic linked library, which is loaded by the *Busola* simulator and used during simulations. Main implemented functionality is listed below:

- Kalman filter—the implementation of this filter is used to estimate the spacecraft inertial state: attitude quaternion, angular rate vector and optionally the parasitic magnetic dipole. Such estimation is always necessary to be done on

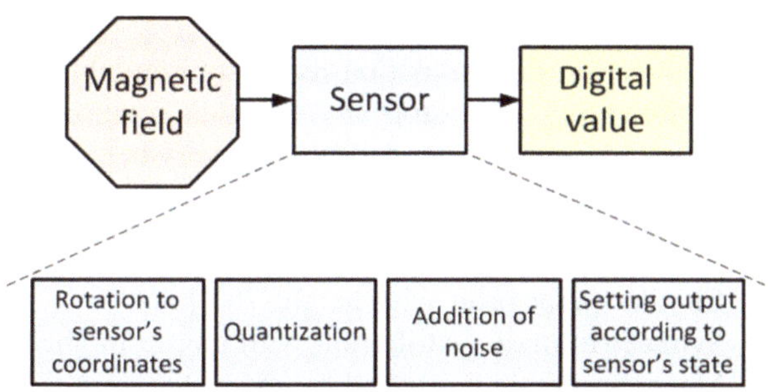

Fig. 4 An example of how the magnetometer sensor is implemented

board of the spacecraft to calculate the control signals, to change in example satellite's orientation,
- Sun position model—this unit calculates the sun position in the ECI (Earth Centered Inertial) orientation frame,
- Magnetic field model—this module calculates the magnetic field,
- Control algorithms—this module provides several control algorithms including: B-dot control, 3-axes stabilization, reaction wheels desaturation etc.,
- Satellite movement on the orbit—this unit implements all laws that describes the movement of an object around the Earth,
- XML—this library allows to access xml files. The files are used to store the current spacecraft configuration.

4.4 Parameters of the Simulator

Parameters of orbit and the satellite itself are configurable, allowing for usage of the tool in different scenarios. Orbit is parameterized with a set of numbers in the commonly used format Two Lines Elements (TLE). Most important data in this set are Keplerian elements and drag coefficient. Position of the satellite on orbit in a given instant of time is calculated by the widely used algorithm Simplified General Perturbation Model (SGP4).

Satellite body is assumed to be a rigid body, so it is characterized by inertia tensor matrix 3×3.

Each of the ADCS components (actuator or sensor) has a set of specific parameters. Common for all of them is relative attitude, or direction of the main axis of the component with respect to the satellite body reference frame. All of the components have parameters describing precision of measurement or actuation. Actuators can act in limited, configurable ranges. Current implementation of the simulator covers: reaction wheels and magnetorquers as actuators, magnetometer, Sun sensor, star tracker and gyro as sensors.

5 Sample Results

In this section the simulation results of detumbling maneuver are shown. Aim of the detumbling is to stop any eventual rotation of the satellite. Most common situation when this is needed is at the very beginning of the satellite's life, when it is released from a launcher.

Parameters of the simulation are typical for the satellite BRITE-PL:

- Circular polar orbit with height 700 km.
- Symmetric mass distribution with inertia moment 0.04 kgm^2.
- Magnetic dipole generated by magnetorquers is limited to 0.3 Am^2.

- Precision of magnetometer 100 nT (1sigma).
- Initial angular rate of satellite rotation was 10 deg/s.
- Time interval of simulation: $\Delta t = 1$ s.

Parameters of other components of the satellite are not relevant, as in the testing case only magnetometer and magnetorquers were used.

For detumbling, the *B-dot* algorithm is used which is described by the control law

$$\boldsymbol{m}_c = -k\dot{\mathbf{b}}, \tag{9}$$

where **m** is a magnetic dipole generated by control coils, $\dot{\mathbf{b}}$ is the time derivative of measured Earth magnetic field taken in the body frame, and k is a controller gain. Simulation results performed for different gains are shown on Fig. 5 and Fig. 6. Results have been obtained for estimated *BRITE-PL* satellite parameters and all disturbances mentioned in Sect. 3 have been taken into account.

Interesting phenomenon can be observed in Figs. 5 and 6. When the gain is large ($k = 1e + 6$) and the control magnetic dipole is not limited, unexpectedly the effectiveness of the detumbling is very poor. It can be explained this way: rotation vector ω follows very strictly the direction of magnetic field B, so the corrections, which are perpendicular to B, almost do not change the magnitude of ω. The situation improves dramatically when the control is limited to 0.3 Am^2 for each of the coils. Similar effectiveness is obtained with smaller gain, $k = 3e + 5$.

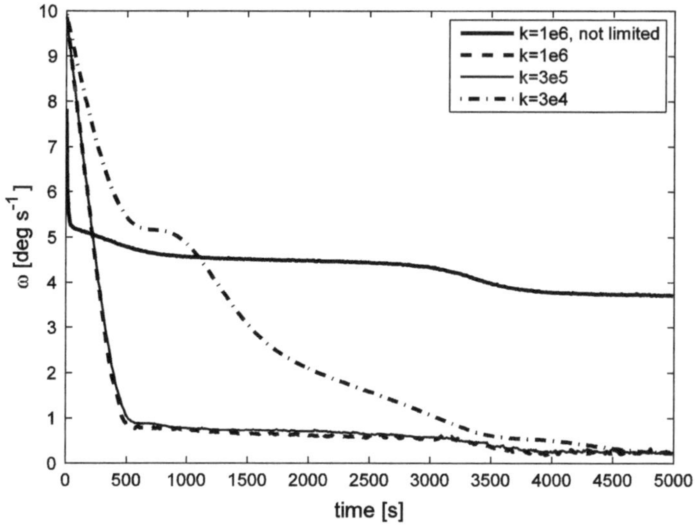

Fig. 5 Evolution of rotational velocity vector magnitude for different values of parameter k. All simulations are performed with coil output limit at 0.3 Am^2 except the first waveform, where no constraints are given

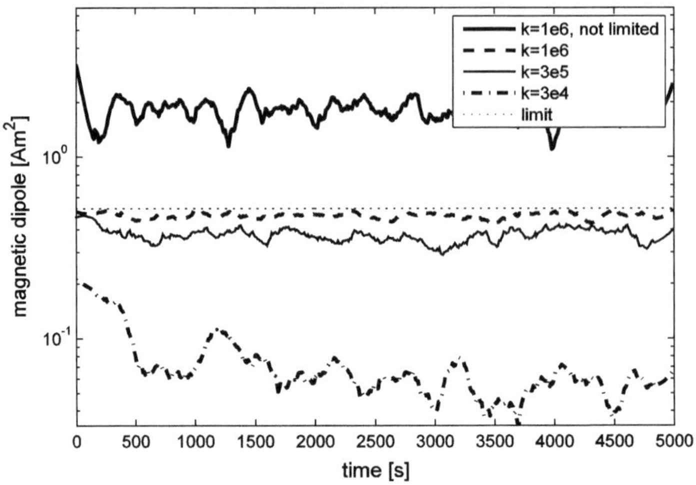

Fig. 6 Evolution of generated magnetic dipole magnitude. Waveforms are smoothed using *Savitzky-Golay* filter. *Dotted horizontal line* represents actuation limit

In this case the control values are smaller in comparison with the case of using stronger gain. Conclusion of the simulation is that k = 3e + 5 is close to optimal.

6 Conclusion

Spacecraft attitude dynamics simulation tool has been developed. It is based on object oriented programming principles implemented in *MATLAB* environment, which adds flexibility to software. The tool gives the opportunity not only to simulate and predict the spacecraft behavior in Earth orbit, but also allows to carry out tests of ADCS software before the satellite launch. Sample simulation results shown in Sect. 5 prove simulator usefulness in tuning the control law gain to get required performance.

The developed ACS software simulation tool can be used in a multitude of scenarios that cover a wide area of space robotics applications: from designing and performing docking maneuvers to practicing satellite formation flying.

Its built-in adaptability can allow easy plug-into another package if needed.

References

Angeles J (1997) Fundamentals of robotic mechanical systems. Springer, London

Baraff D (2001) Physically based modeling: rigid body simulation, Siggraph 2001 Course Notes. http://graphics.cs.cmu.edu/courses/15-869-F08/lec/14/notesg.pdf

Eberly D (2010) Quaternion algebra and calculus. http://www.geometrictools.com/. Accessed 15 Oct 2011

Fredrickson SE (2001) Mini AERCam: a fee-flying robot for space inspection. In: ISA 2001 technical conference. http://ntrs.nasa.gov/archive/nasa/casi.ntrs.nasa.gov/20100033358_2010035225.pdf

Hughes P (2004) Spacecraft attitude dynamics. Dover Publications Inc, Mineola

Moffat AFJ, Weiss WW, Rucinski SM, Zee RE, van Kerkwijk MH, Mochnacki SW, Matthews JM, Percy JR, Ceravolo P, Grant CC (2006) The Canadian BRITE NanoSatellite mission. In: Proceedings of ASTRO 2006—13th CASI (Canadian aeronautics and space institute) Canadian astronautics conference, Montreal, 25–27 April 2006

Murray RM, Li Z, Sastry SS (1994) A mathematical introduction to robotic manipulation. CRC Press, Boca Raton

Orleański P, Graczyk R, Rataj M, Schwarzenberg-Czerny A, Zawistowski T, Zee RE (2010) Proceedings of SPIE 7745, 77450A. doi:10.1117/12.872151

Pervin E, Webb JA (1983) Quaternions in computer vision and robotics. In: Proceedings of the international conference vision and pattern recognition, Washington

Schaub H, Junkins JL (2003) Analytical mechanics of space systems. AIAA Education Series, Reston

Trinidad K, Humphries K (2005) Nasa successfully demonstrates innovative nanosatellite system. http://www.nasa.gov/home/hqnews/2005/jun/HQ_05154_nanosatellite.html

Vicci L (2001) Quaternions and rotations in 3-space: the algebra and its geometric interpretation. TR01-014, Department of Computer Science, University of North Carolina, Chapel Hill

Laser Photography in Selective Space Imaging and Navigation

Marek Piszczek and Marcin Kowalski

Abstract The aim of this chapter is to discuss the properties of cameras using lasers as an illuminator during the image acquisition process. The basic types of ToF (time-of-flight) cameras are characterized, with the results of a PMD Cam-Cube 3.0 ToF camera, and a laboratory laser photography device test presented. This will be applied in showing and analyzing the application capabilities of laser cameras. The unique qualities of ToF cameras allow them to be a valuable tool in the observation-measurement process, and they can be applied in outer space explorations and space navigation amongst other industries.

1 ToF Camera Types

Every imaging information system can be characterized by its parameters. The ability to describe shape, size and/or spatial location of objects, and physical phenomena, distinguish the potential ability of photogrammetric information solutions. The information capacity of a system is dependent on a variety of acquisition module parameters, such as distributing ability, angular field of view, knowledge of internal and external orientation elements of remote sensing, and analytics modules that are responsible for data processing and analysis. The complexity of imaging information is segregated into four different types of resolutions for imaging systems: spatial, time, radiometric and spectral.

In order to describe the above-mentioned laser cameras in accordance to the data, one should know that laser cameras, similarly to other cameras can be characterized by different radiometric resolutions. This is a result of detectors that

M. Piszczek (✉) · M. Kowalski
Institute of Optoelectronics, Military University of Technology,
2 S. Kaliski Street, Warsaw, Poland
e-mail: mpiszczek@wat.edu.pl

J. Sąsiadek (ed.), *Aerospace Robotics*, GeoPlanet: Earth and Planetary Sciences,
DOI: 10.1007/978-3-642-34020-8_4, © Springer-Verlag Berlin Heidelberg 2013

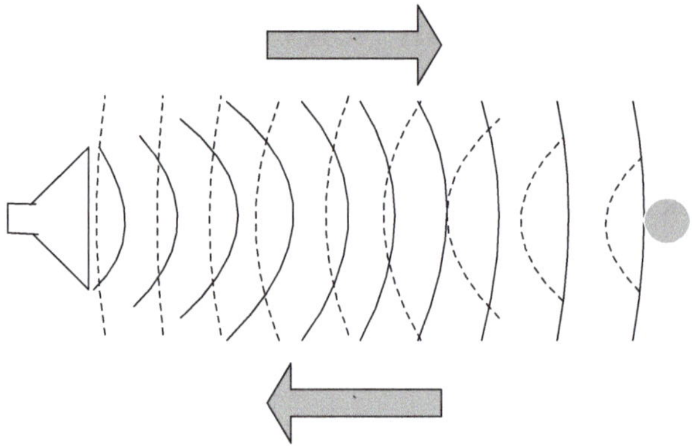

Fig. 1 Illustration of a ToF camera principle

are in use in these cameras. The radiometric resolution is, however, not the principal parameter of the above mentioned imaging devices.

The ToF (Time-of-flight) camera is an imaging device that measures the time of flight of an electromagnetic wave between an illuminator, a lens and a camera (Fig. 1) to create an image; therefore, it is applying the usage of active scene illumination.

What distinguishes ToF camera from stereoscopic camera is that the first one uses single detection unit (detector array and a lens) 3D image acquisition instead of two units in stereo cameras. The above-mentioned data results in some differences in the interpretation of spatial and time resolution. In traditional photography we are able to determine the angular range of observation, i.e., every single pixel of an image can be matched with its angular field of view (two-dimensional information). However, the information that laser cameras store in every single pixel is three dimensional, not an angular quantity (Fig. 2).

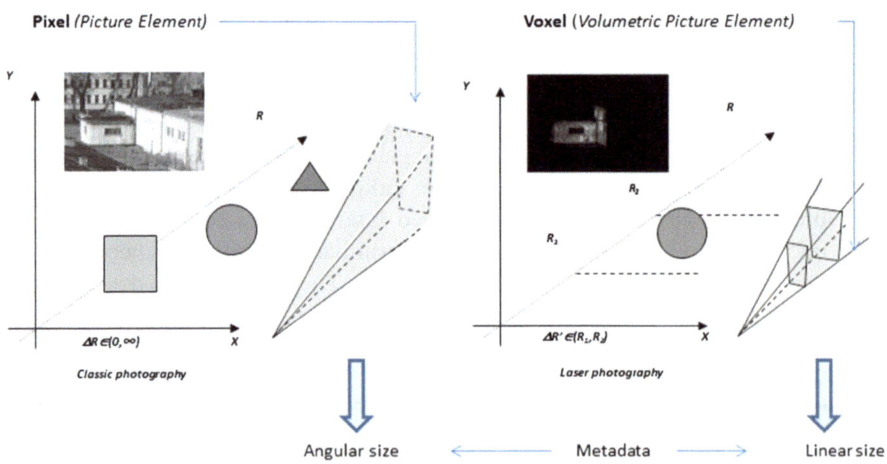

Fig. 2 Spatial properties of imaging

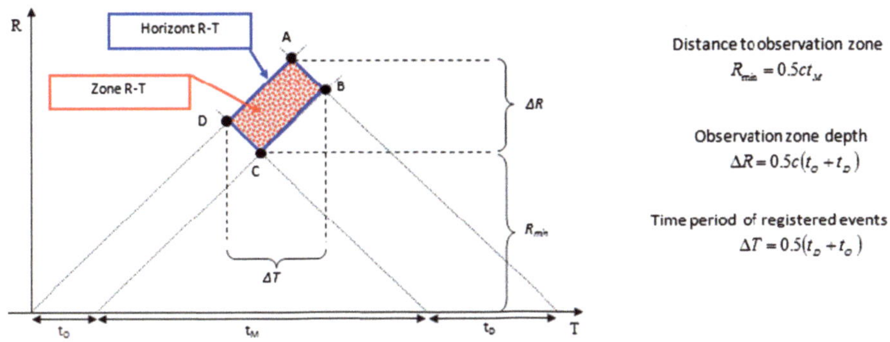

Fig. 3 An idea of building time-spatial frame

A notion of time resolution has similar properties to angular resolution. Laser cameras working in a single image acquisition mode are able to register very fast changeable phenomena, particularly in cases relative to very narrow time acquisition and spatial range events. This range can be modified using the time-spatial framing method, as illustrated in Fig. 3.

ToF cameras using laser illumination acquire images in very thin lines of the electromagnetic spectrum. Due to a narrow spectrum, the sun light illumination impact is minimized (in ToF cameras, natural light is a parasitical radiation). In order to enhance the image contrast, radiation acquired by the ToF camera should be filtered as to reduce the impact of sunlight. Filtering can be described using Eq. 1, where parameter R stands for spatial radiation properties, I for the radiometric and spectral properties, I_A for laser lighting, and I_B is a component of sun light which is filtered $\Delta\lambda$.

$$I_L \infty \int_0^{\infty} \int_{\Delta\lambda} I_A(R,\lambda) dR d\lambda + \int_{\Delta R} \int_{\Delta\lambda} I_B(R,\lambda) dR d\lambda \tag{1}$$

According to the light source and data acquisition method used, ToF cameras can be divided into three types:

a. Impulse light source with digital time counters,
b. Modulated light source with phase detectors,
c. Impulse light source with gated acquisition time.

The first solution using impulse light sources with digital time counters is based on a specialized sensor. This sensor is an array of detectors working like an impulse range-finder. An example of these solutions is illustrated in Fig. 4 (Stettner et al. 2001).

Similar ideas were applied in the second type of ToF camera—a solution with modulated light source with phase detector, where elementary detectors work like phase range-finders. The solutions of this kind are becoming more common in

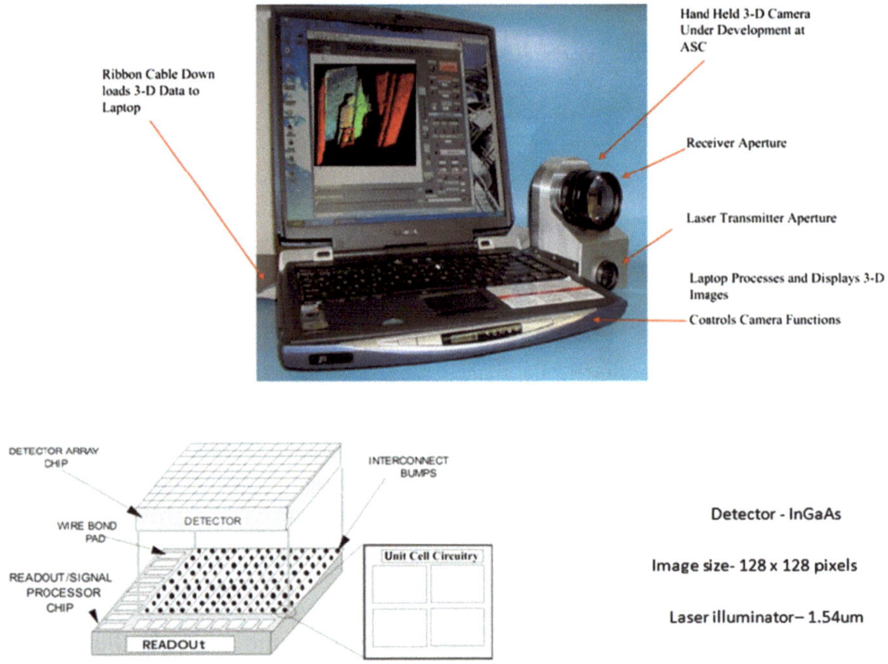

Fig. 4 ToF camera with impulse light source and digital time counters

commercial markets, with multinational-companies like PMD Technologies, SoftKinetic or Panasonic involved in the production of such cameras. An example of the SwissRanger cameras made by MESA Imaging is shown in Fig. 5.

The last type of ToF cameras is a camera consisting of an impulse light source with a gated acquisition time. These forms of cameras are referred to as RGI

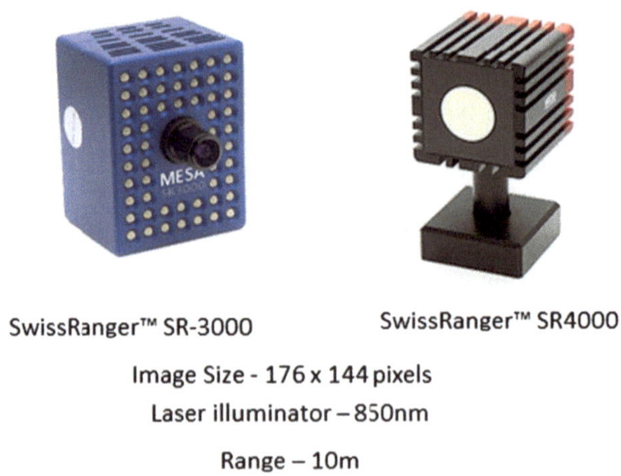

Fig. 5 ToF camera with modulated light source and phase detectors

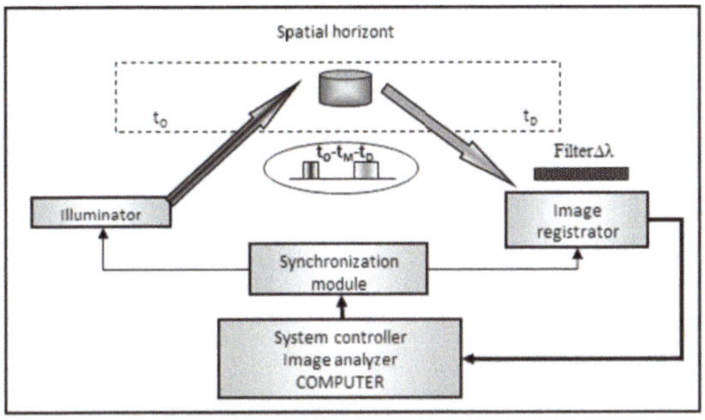

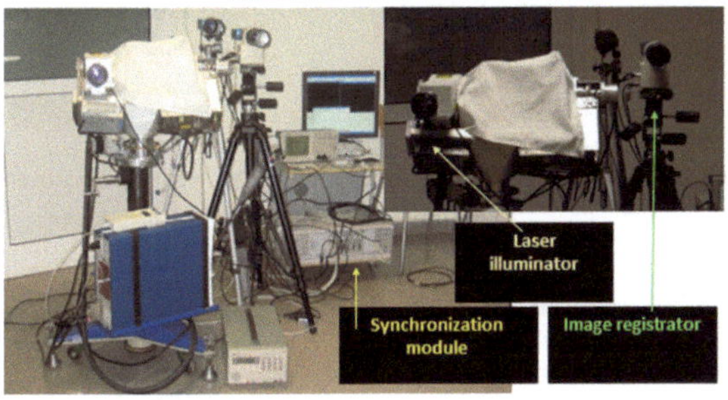

Laser illuminator	Laser NdYAG, energy10 mJ. λ = 530nm (second harmonic, illumination time t_O = (10-20) ns
Image registrator	camera HSC-145BW image amplification MCP detection timet_D = (10-70) ns
Synchronizati on module	Sequencer sterujący sekwencjami czasowymi z krokiem Δt = 5ns

Fig. 6 ToF camera with impulse light source with gated acquisition time

(Range-Gated-Imaging) cameras (McDonald et al. 1999; Andersson 2006), and work according to the time-spatial framing method. An example of the laboratory measurement system developed in the Institute of Optoelectronics, WAT, is shown in Fig. 6 (Piszczek et al. 2007a, b).

2 Spatial Measurements with ToF Cameras

The aim of camera tests was to define proper camera work conditions and verify the impact of camera parameters to acquired data.

Spatial images acquired using ToF camera can be divided into zones. Figure 7b shows a laboratory stand divided into few distance zones—the nearest zone is marked with red colour and the furthest is marked with blue colour (Figs. 8, 9).

a. Influence of illumination time on registered images (Table 1).

3 ToF Cameras Functionalities

The basic properties and functionalities of laser cameras are:

a. Acquisition and visualization of spatial information
   Information about spatial location of observed objects is not synthesizable as a result of complicated analysis. Laser cameras have a unique property that

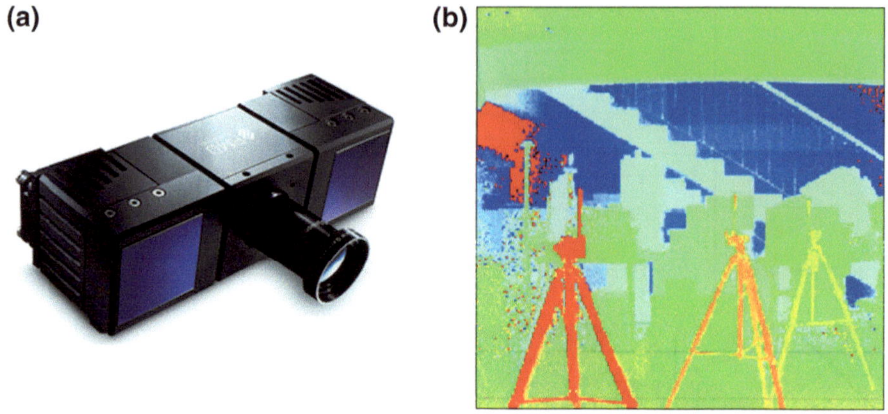

Fig. 7 **a** ToF camera: PMD[vision]® CamCube 3.0. **b** Spatial image acquired with ToF camera

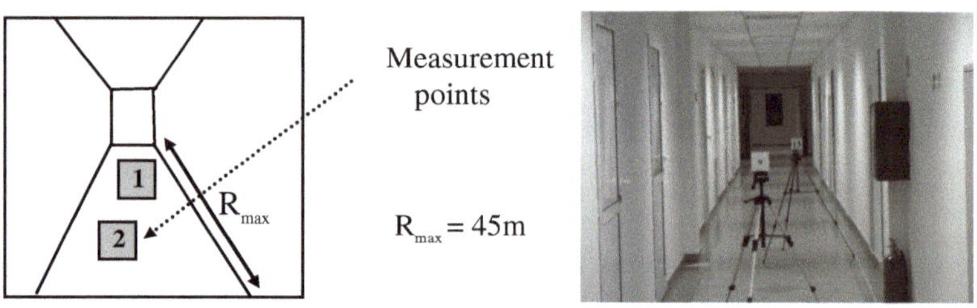

Fig. 8 Laboratory measurement space

Table 1 Parameters of the recording of images—test the time of illumination

Frame	t_i (ns)	t_M (ns)	t_D (ns)	R_{min} (m)	ΔR (m)
a)	10	65	65	9.7	11.2
b)	20	55	65	8.2	12.7

allows them to acquire and expose spatial information. Cameras equipped with time counters and phase detectors deliver spatial information as a map of depth (Fig. 10a) (Schuon et al. 2008; Medina et al. 2006; Medina 1992). This map can be displayed as a coloured or monochromatic map, hence image resolution in imaging plane can be even 0.05 Mpix. Cameras with time gating give images presenting precisely defined fragments of space (distance, depth of observation). Additionally we acquire information about the reflection coefficient (Fig. 10b) hidden in a monochromatic map of depth. Therefore, images are more realistic and have a bigger resolution in its imaging plane up to 1 Mpix.

b. Spatial selectivity of imaging

On top of giving spatial information, ToF cameras can make spatial selection of observed objects. Time counting or phase analyzing in acquisition process cameras produce continuous space information. In this type of ToF camera the

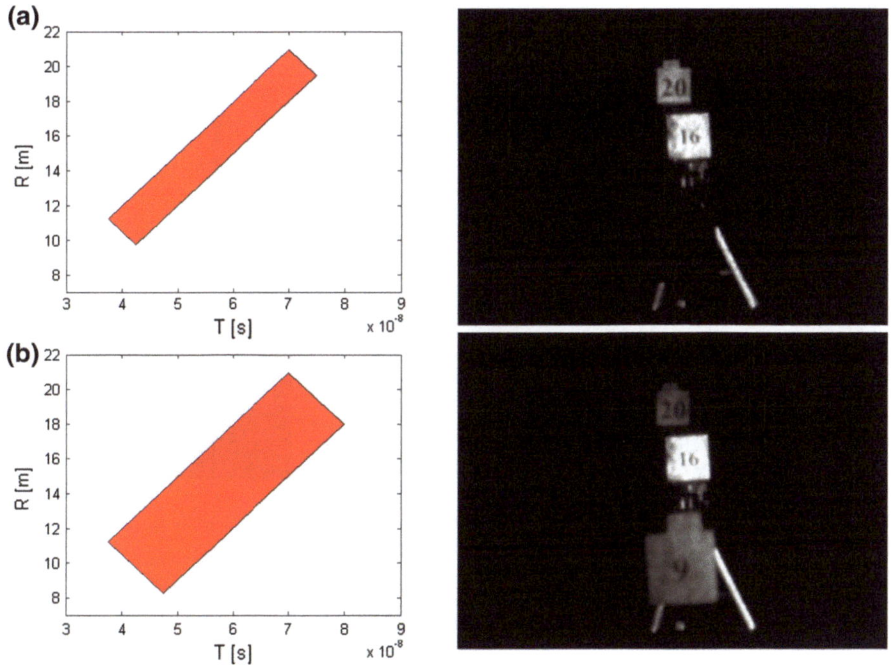

Fig. 9 Results of testing the influence of illumination time **a** $t_0 = 10$ ns, **b** $t_0 = 20$ ns

selection of fragments of the observed scene is derived using a special application. Cameras with time gating capabilities, unlike the time counting of phase analyzing cameras, can select a part of an observed scene during the acquisition process. However, to build full spatial profile, many scan of space are needed (Fig. 11).

c. Image quality enhancement
   One of the most important needs of recent vision systems is to build an observation system that can deliver good quality images regardless of lighting conditions (day, night)and the presence of dissipative factors (fog, rain, snow) appearing between a lens and an observed object (Piszczek at al. 2006, 2008). As such, due to spatial selectivity there is a possibility to minimize the impact of unbeneficial atmospheric phenomena, as shown in Fig. 12.
d. Image processing–autosegmentation
   Another valuable property of ToF cameras is the ability to obtain image processing during data acquisition. Imaging data processing is however, substantially limited in scene segmentation and object extraction. Laser cameras are able to deliver segmented image as an effect of spatial selectivity. Image segmentation is mostly a post-processing method; however, laser cameras offer this capability during the acquisition process (pre-processing). Autosegmentation is a post-effect of spatial selectivity, as illustrated in Fig. 13.

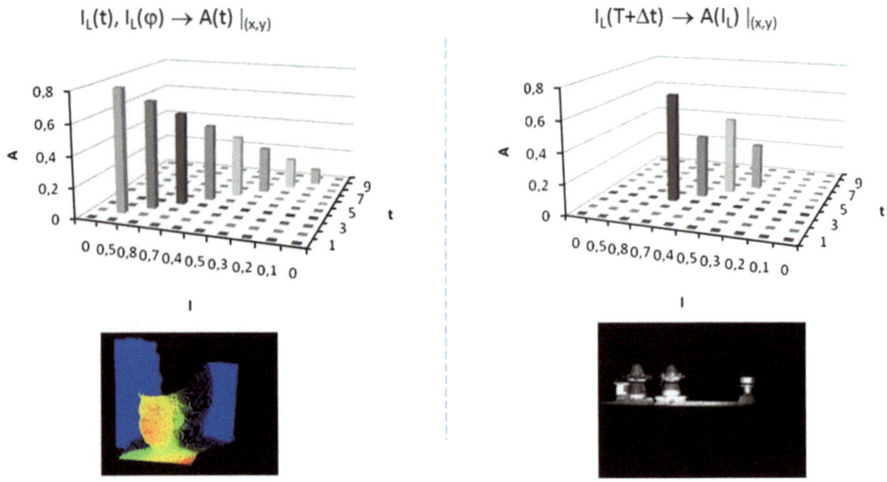

Fig. 10 Acquisition and visualization of spatial information **a** camera with time counters and phase detectors, **b** cameras with time gating

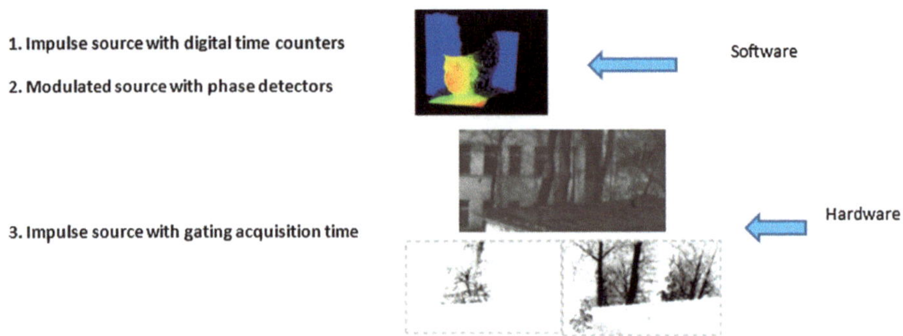

Fig. 11 Spatial selectivity of ToF cameras

e. Image photogrammetric analysis
   Metadata delivered by images from a device using the time-spatial framing method can be used to make an advanced photogrammetric analysis. An image acquired using ToF camera connected with the orientation parameters of acquisition modules offer us the capability of defining linear object dimensions (Fig. 14a). Spatial information can be visualized (Fig. 14b) or analyzed thanks to autosegmentation (Fig. 14c).
f. Spatial modeling
   Information about the internal and external orientation of image acquisition modules can be very useful during analysis. This information can be used in the spatial modeling process based on data collected from single measurement points. The simplest form of 3D visualization is a laminar model (Fig. 15).

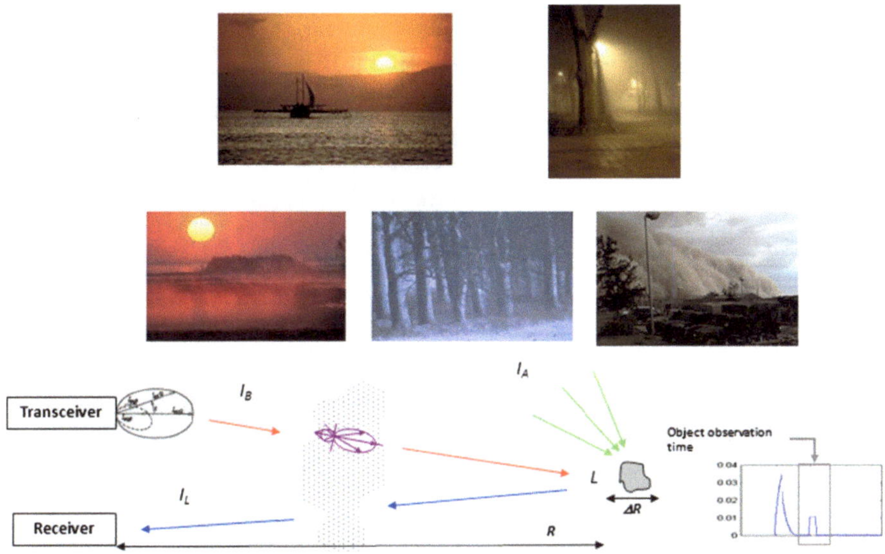

Fig. 12 Idea of improving image quality by using time-spatial framing method

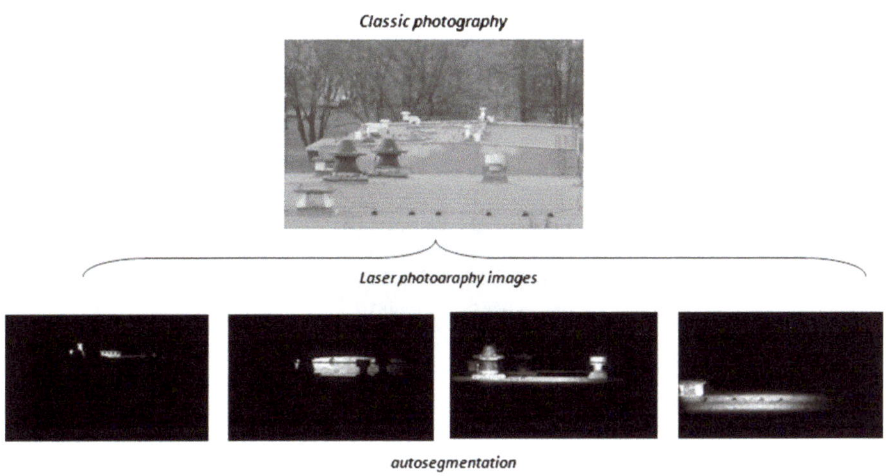

Fig. 13 Autosegmentation of observed scene using ToF camera

The composition of two ToF camera images, taken from different observation points, can offer a potential prospect to building realistic 3D images. This composition can also be used to render scene objects with imaging material from laser cameras.

All of the above mentioned properties show that ToF cameras can be used as an advanced navigation tool. Traditional navigation systems based on GPS,

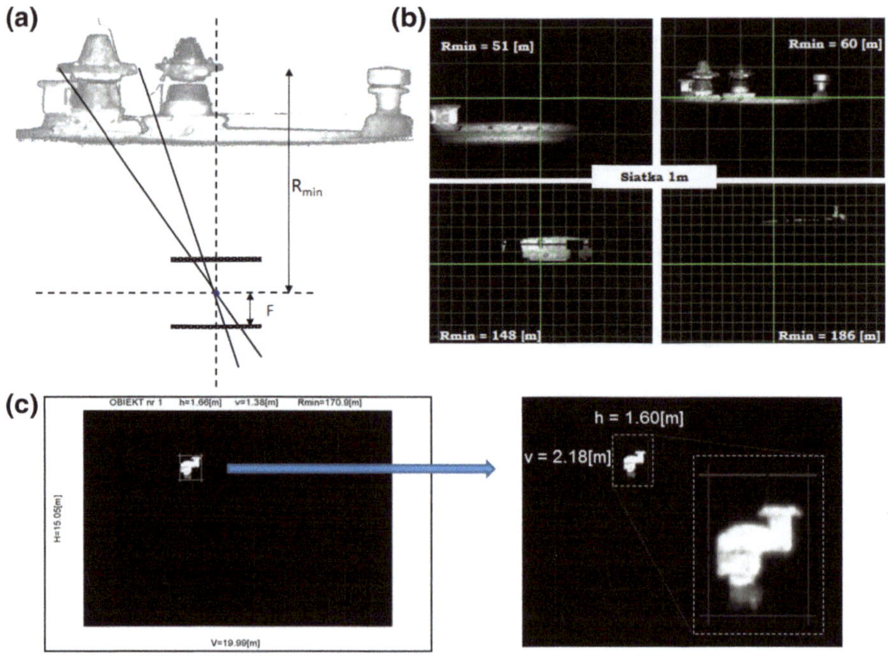

Fig. 14 Image photogrammetric analysis **a** visual valuation, **b** automatic scene analysis, **c** autosegmentation

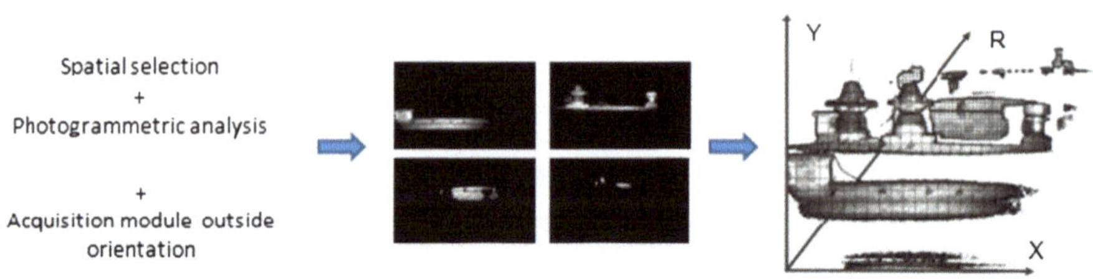

Fig. 15 Laminar model acquired by range-gated-imaging camera

spatial orientation systems or inertial navigation system need ground-based and cosmic infrastructure, gravitation and magnetic fields. However, these factors are not widely available. Assuming that orientation points are available, a local positioning system can be developed using the method of spatial modeling. Furthermore, imaging data from ToF camera can be used to detect and identify orientation points during the navigation process. Additional spatial information given by laser cameras facilitates defining the position of the camera in space relative to orientation points. This process is realized using triangulation analysis. The general basis behind laser camera navigation is illustrated on Fig. 16.

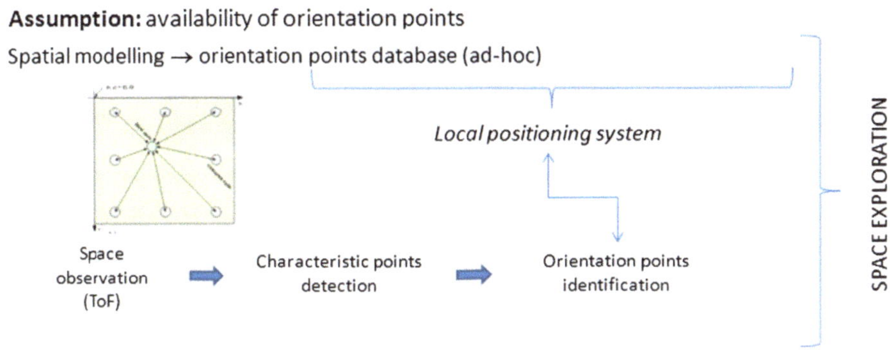

Fig. 16 Idea of navigation realized in local reference system using laser camera

ToF camera navigation can be employed in a variety of ways. It can be applied in widespread areas and directly with the environment. For example, a mobile platform equipped with a ToF camera can simplify auto navigation in limited spaces due to the detection of mobile or forbidden zones (Fig. 17).

The proposed navigation solutions can be utilized to explore any possible space. The basic element of the described spatial information system is a ToF camera (laser camera).

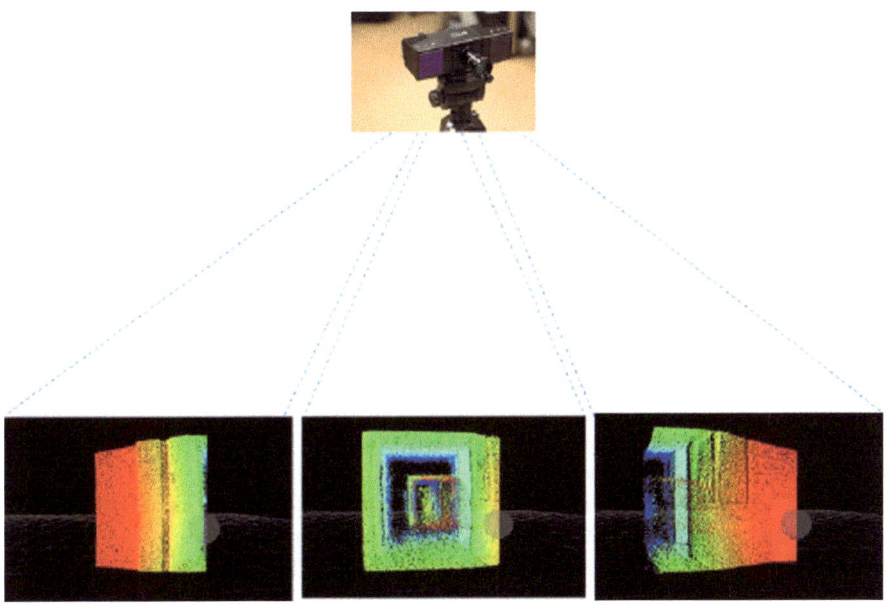

Fig. 17 Mobile and forbidden zones for autonavigation of mobile platform in limited space

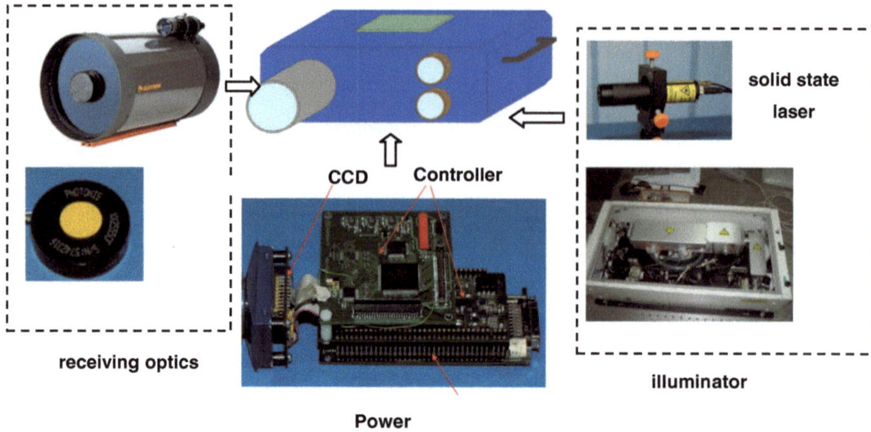

Fig. 18 Laser photography device

4 Hardware Configuration

A model of the Laser Photography Device (LPD) built the Institute of Optoelectronics, WAT, offers a 1 ns time resolution which is equal to about 1 m spatial resolution. The Laser Photography Device is a reconfigurable device. It uses two types of illuminators:

a. Nd:YAG laser (532 nm) with 5 ns pulse (380 mJ)
b. Solid state laser—850 nm (100 W) and 905 nm (220 W) with a pulse time of several to tens of ns.

1 megapixel sensor with 1 ns shutter (achieved by using MCP image amplifier) is a receiver module used in the camera.

The illuminator is not the only reconfigurable element of the LPD; the camera optics is also reconfigurable. The biggest lens (used with Nd:YAG illuminator) offers a focal length f = 2,800 mm with F = 1/10. The hardware configuration of the Laser Photography Device is shown in Fig. 18.

There are two types of the LPD system configuration (Fig. 19):

- ver. A—stationary configuration—used with Nd:YAG laser. The whole device is placed on a positioning platform equipped with a set of sensors determining location and spatial orientation. Device control is realized by a central information unit (a system controller).
- ver. B—mobile configuration—equipped with solid state lasers. The device is handy and can be operated manually by a user. Images acquired by a mobile version of the LPD are presented directly on the LPD's display.

Because research on the Laser Photography System is a work on a novel device and system, the following software solution is proposed:

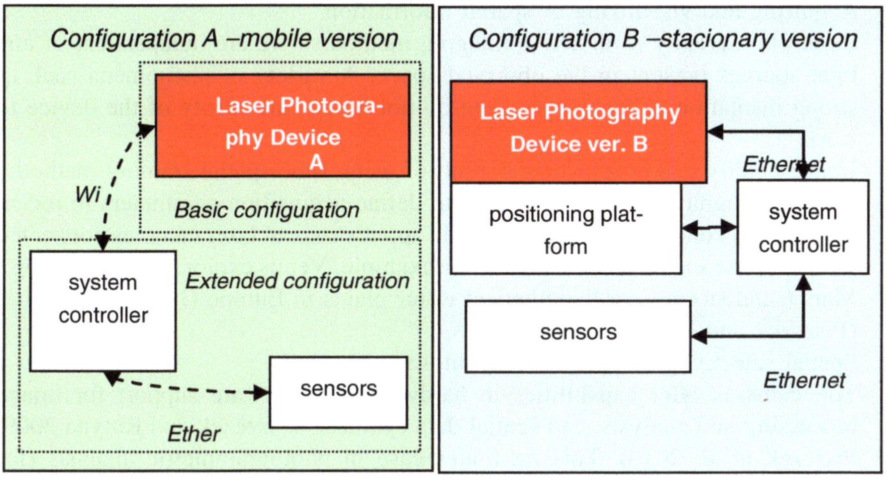

Fig. 19 Basic LPD configurations

a. research software–its task is to analyze the LPD work parameters and various system configurations,
b. system's software–is the synthesis of different research solutions realizing specific system functions for various system configurations.

Further development of the Laser Photography Device in the area of improvement of camera parameters and adding new features seems to be justified. At this stage of development it can be said that using laser illumination can improve the process of defining acquisition parameters (time parameters as well as energetic).

5 Summary

Unique and characteristic properties of ToF cameras are features of this particular type of imaging device. The functionalities of ToF cameras can be very useful and essential in the application of space information systems. Imaging information aspects connected with laser cameras should particularly be highlighted:

a. Spatial—ability to define 3D observation area,
b. Time—capability to register dynamic phenomena (relativistic speed),
c. Radiometric—ability to register and synthesize images with wide range of dynamics (thanks to illumination levels and amplification controlling)
d. Spectral—imaging data acquisition in very thin spectral range.

The possibility of using ToF cameras in space exploration is connected with the following functionalities:

a. Acquiring and visualizing of spatial information
Laser photography is an active imaging method so we are independent of any light sources present in the observed space. A variety of phenomena such as strong insolation or lack of sunlight do not hinder the ability of the device to work.
b. Image quality enhancement is a result of using time-spatial framing method.
Laser photography offers the ability to define acquisition parameters to recent observation conditions. One of possible applications of laser cameras connected with universe explorations may be, for example, Venus exploration (dense fog), Mars (sand storms), or hypothetical water plants in Europe (Jupiter's satellite) (Piszczek and Rutyna 2008).
c. Spatial selectivity of imaging and autosegmentation
ToF cameras offer capabilities in hardware and software support for image processing and analysis, and spatial data synthesis (Piszczek and Rutyna 2009; Piszczek et al. 2010). This facilitates ease in photogrammetric analysis (for navigation and spatial processing assistance).

Acknowledgments The special capabilities of imaging information acquisition found in laser cameras creates a very valuable observation-measurement tool. Research about time-spatial framing method started in the Institute of Optoelectronics, WAT in 2005 and recent research realized during project OR00000312 is financed by National Research and Development Center.

References

Andersson P (2006) Long-range three-dimensional imaging using range-gated laser radar images. Opt Eng 45:034301

McDonald TE Jr, Yates GJ, Cverna FH, Gallegos RA, Jaramillo SA, Numkena DM, Payton JR (1999) Range-gated imaging experiments using gated intensifiers. Proc SPIE 3642:142

Medina A, Gayá F, and Pozo F (2006) Compact laser radar and three-dimensional camera. J Opt Soc Am A 23:800–805

Medina A (1992, 2008) Three dimensional camera and rangefinder. United States Patent 5081530 Piszczek M, Rutyna K, Pozyskiwanie informacji przestrzennej z wykorzystaniem przestrzenno-czasowego kadrowania obrazów, (Acquisition of spatial information with the use of time-space framing method) Roczniki geomatyki 2008—Tom VI, Zeszyt, vol 7, pp 73–84 (in Polish)

Piszczek M, Rutyna K (2009) Symulacje komputerowe procesu selektywnego zobrazowania przestrzeni z uwzględnieniem różnych warunków atmosferycznych i klimatycznych (computer simulations of selective imagning of space in different atmospheric and climatic conditions), Elektronika nr 2/2009, pp 31–36 (in Polish)

Piszczek M, Rutyna K, Szustakowski M (2008) Imaging of space with using optoelectronic observation system with active illumination. Eur Phys J Spec Top 154(1):153–157

Piszczek M, Rutyna K, Szustakowski M (2006) Koncepcja systemu fotografii laserowej do identyfikacji obiektów w warunkach słabego oświetlenia, (concept of laser photography system for object identification under weak illumination conditions) IX Konferencja Naukowa COE-2006, str. 51–55, Materiały konferencyjne COE 2006, Kraków—Zakopane 19-22 czerwca 2006, ISBN 8388309315, pp 51–55 (in Polish)

Piszczek M, Rutyna K, Szustakowski M (2007a) Optoelectronic observation system with active illumination. International congress on optics and optoelectronics 2007. SPIE Eur Opt Optoelectron, vol 6585
Piszczek M, Rutyna K, Szustakowski M (2007b) Rejestracja obrazów metodą kadrowania czasoprzestrzennego (recording of images by using the time-space framing method), Pomiary Automatyka Kontrola, vol 53, nr 9bis/2007, pp 531–534 (in Polish)
Piszczek M, Zagrajek P, Ryniec R (2010) Pozyskiwanie i udostępnianie informacji obrazowej w systemach rozproszonych, (acquisition and availability of imagne information in scattered systems), Zeszyty naukowe ETI Politechnik Gdańskiej, Tom 19/2010, pp 47–52, (in Polish)
Schuon S, Theobalt C, Davis J, Thrun S (2008-07-15) High-quality scanning using time-of-flight depth superresolution. Written at Anchorage, Alaska. IEEE computer society conference on computer vision and pattern recognition workshops
Stettner R, Bailey H, Richmond RD (2001) Eye-safe laser radar 3D imaging. Proc SPIE 4377:46

The Project of a Simple Drive for CCD Observation Cameras

Ewa Knap, Jerzy Grygorczuk and Paweł Pyrzanowski

Abstract Curiosity is a passport to the Universe. The progress in technology made admiring the nightly sky easier than ever. A mechanism utilizing two CCD cameras based on equatorial mount is a simple design characterized by high precision. The design had two basic assumptions. The first concerned the accuracy of object positioning: 6 s of arc, second related to the kinematics of the system—continuity of operations and the possibility to switch to a different object of observation in a few seconds. All components used in this design are available on the market. Never before has technology given so many tools for taking a look into the Cosmos. Though the presented mechanism might seem simple, amateur even, it can only be treated as one of its many advantages.

1 Introduction

"What could be more beautiful than the sky embracing all that is beautiful"—the words said almost five centuries ago by Mikolaj Kopernik. Despite the passing of time they did not lose any of their timeliness. The development of new technologies made marveling at the beauty of the nightly sky easier than ever. More and more technologically advanced mechanisms allow zooming in on almost every

E. Knap (✉)
The Faculty of Power and Aeronautical Engineering,
Warsaw University of Technology, Nowowiejska 24,
00-665 Warsaw, Poland
e-mail: ewaknap@poczta.onet.pl

J. Grygorczuk
Space Research Centre of the Polish Academy of Sciences,
Bartycka 18A, 00-716 Warsaw, Poland

P. Pyrzanowski
Institute of Aeronautics and Applied Mechanics,
Warsaw University of Technology, Nowowiejska 24,
00-665 Warsaw, Poland

J. Sąsiadek (ed.), *Aerospace Robotics*, GeoPlanet: Earth and Planetary Sciences,
DOI: 10.1007/978-3-642-34020-8_5,

nook of the Universe. Thanks to devices placed all around Earth, including its orbits, it is possible to retrieve images of amazing quality, which could have only been dreamed of in the past. At this very moment planetary rovers are exploring the surface of Mars, nearly 3,000 satellites are observing the Earth and Cosmos, space probes are discovering the secrets of planets and moons as well as investigating the mysteries of distant galaxies.

The presented mechanism may be used as one of tools for observing the Universe from the Earth's surface, placed in a research station as well as, in somewhat changed form, scientific equipment of a satellite.

This concept of precise astronomic devices based on equatorial mount is not new. One of those was Pi of the Sky, which was a big inspiration for this simple drive. Pi of the Sky was made for investigating short timescale astrophysical processes with main focus on Gamma Ray Bursts (GRBs). The apparatus is installed in Warsaw University Observatory at Las Campanas Observatory in the Atacama desert in Chile.

2 The Project of a Drive for Two CCD Observation Cameras

2.1 Main Assumptions

The project of a simple drive for CCD observation cameras with two degrees of freedom had a couple of preliminary assumptions. The first concerned the accuracy of object positioning—initial value was about 6 as. Considering how non-advanced and amateur the project was, this precision of angle reading could be difficult to achieve. However, through application of precise high-quality devices it became possible. The next requirement was assuring the ease of control (by applying, popular in this type of mechanisms, uncomplicated equatorial mount) and smoothness of motions. The whole structure had to be imposed with high rigidity demands—over-constrained construction will not generate significant measurement errors. Moreover, small amount of components, ease of their production and availability on market make this design easy to assemble, reliable and inexpensive.

The mechanism, however, is not enough. Creation of appropriate software, which would use the catalogue of sky maps, is the key aspect of the project. However, this article will only concern the drive of the cameras.

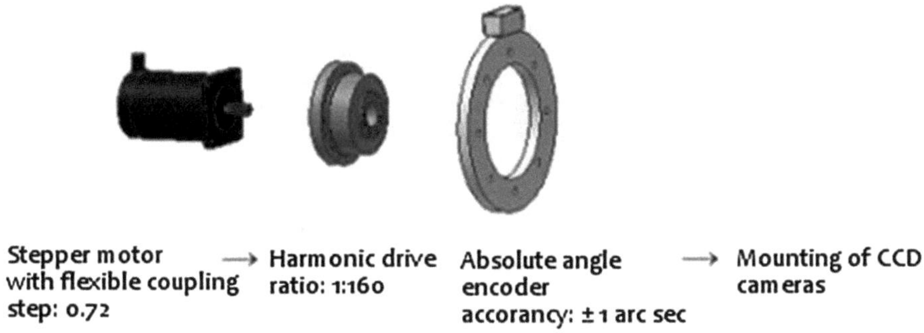

Fig. 1 The kinematic scheme

2.2 Components

The preliminary design of the mechanism is composed of three fundamental modules. The first of these is the structure itself. Its elements are: fork, frame and plate. The fork is a rigid element welded from a rectangle-profiled tube. The frame, to which the whole construction is mounted, is composed of three elements. Its parts are laser-welded to each other. It is jointed with bolts to a rectangular plate. The second module is the camera mounting unit. Two CCD cameras are fixed on a machined plate embedded in two high-precision bearings. The third part is the drive system of declination and ascension axes.

However, the very first idea was different. One of the concepts assumed using epicyclic gearing, other fluid coupling, some ideas considered use different types of motor. But after a market research, it turned out that the drive could be based only on two stepper motors (one per axis) without any other components.

Figure 1 displays the kinematic scheme of the device. The basis of the drive system are two electric stepper motors. The torque is transmitted through flexible coupling to a harmonic drive. The structure upon which the observation cameras are mounted is set into motion by the harmonic drive. An extremely important element is the absolute encoder. It is a part of the feedback loop, which is so significant due to high precision.

All components used in the preliminary design were chosen from catalogues of products currently available on the market.

2.2.1 Motor

Two-phase hybrid stepper motor ZSS 56 produced by Phytron[1] is the main drive of the system. According to catalogue data its holding torque is equal to 700 mNm, detent torque 50 mNm and the rotor's mass inertia—0.24×10^{-4} kg $\times$ m^2.

---

[1] http://www.phytron.com/

Table 1 The data of CGS 32 harmonic drive

Size	Ratio	Limit for repeated peak torque	Limit of average torque	Rated torque at rated speed 2,000 rpm	Limit for momentary peak torque	Moment of inertia	Weight
32	160	484 Nm	281 Nm	178 Nm	892 Nm	1.69×10^{-4} kg $\times$ m^2	0.89 kg

This particular model has been chosen due to its small basic step value of 0.720 (which means 500 steps for full rotation) as well as the high value of maximum rounds per minute (up to 6,000). What is very important is that the producer assures the motor has the ability of micro-step control without the risk of losing any of the smoothness of movement. Such small step divided into 8 parts guarantees extremely high resolution of the motor output. High rpm determines a several second time of changing of the observation object.

Detailed characteristics of the output power and frequency of the motor are to be found on producer's website.

2.2.2 Harmonic Gear

Harmonic gears of German manufacturer "Harmonic Drive[2]" from CSG-2A series have a beneficial power to weight ratio and physical dimensions; what is more, they have high gear ratio (even up to 1:160). These are the main reasons why this model has been chosen. It also has other advantages, like high torque capacity, compact design, high torsional stiffness, zero backlash, high efficiency and simple installation.

The Table 1 below shows the basic characteristics of CSG-2A 32 gear. More information about this series of gears can be found on manufacturer's website.

2.2.3 Encoder

Ultra-high accuracy absolute angle encoder from Rexa series by Renishaw[3] is the most important element ensuring the precision of the mechanism. It guarantees measurements with accuracy of up to 1 arcsecond. As a non-contact encoder, Rexa offers dynamic performance advantages eliminating coupling losses, oscillation, shaft torsion and other hysteresis errors which plague enclosed encoders. Eccentricity error can be easily removed by using two readheads and combining the signals inside the host controller. Maximum rpm of the chosen size (104 mm in diameter) is 4,400 rev/min. More information and performance data are on producer's website.

[2] http://www/harmonicdrive.de/

[3] http://www.renishaw.com.pl/

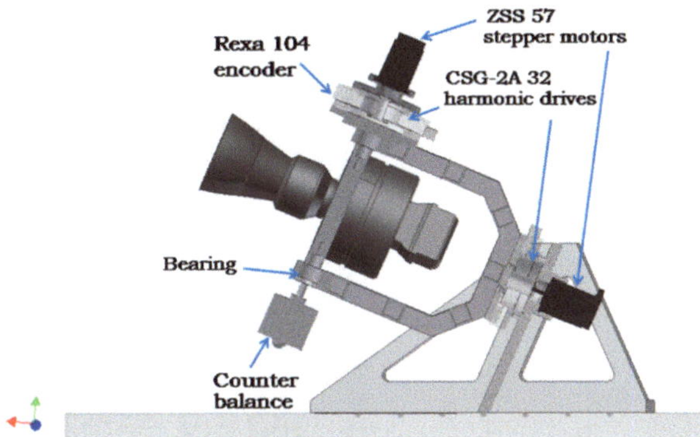

Fig. 2 The cross-section of assembly

Application of such a sensitive encoder allows for high resolution and guarantees fulfilling of the accuracy requirement of 6 s of arc mentioned before (achieved final value of resolution is even greater than the assumed one). Moreover, with properly scripted control program automatic, on-the-run error correction is possible.

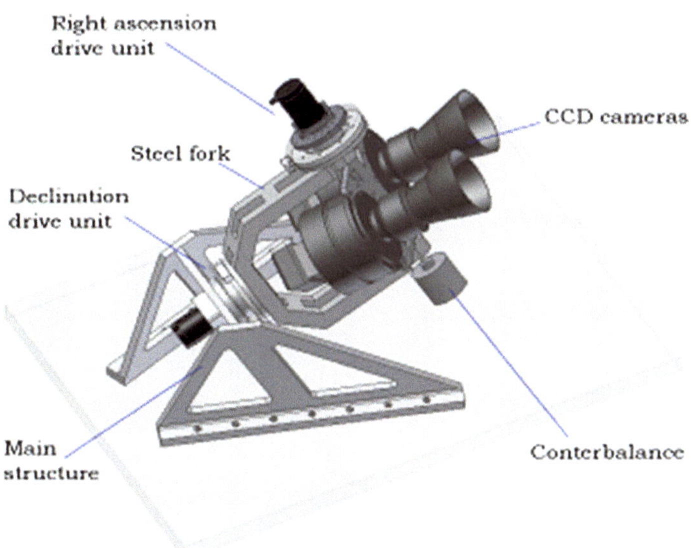

Fig. 3 The structure of project of drive for CCD cameras

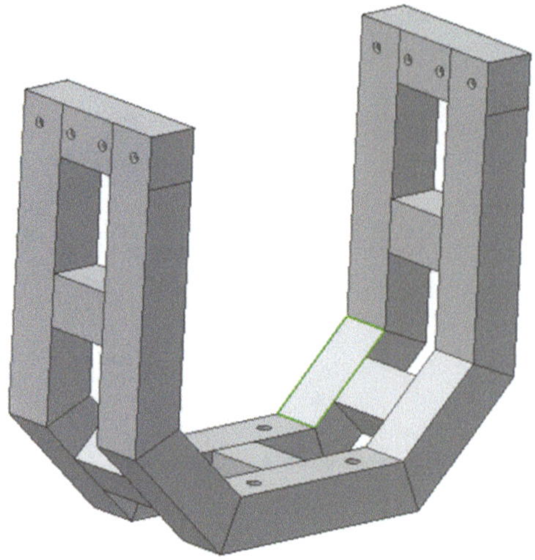

Fig. 4 The fork

2.3 Structure

The main elements of the structure are the bases attached to a table on which the whole construction of fork and camera mountings are fixed Fig. 2. Components are laser-welded to each other, which ensures better exactness and endurance of the joints.

It should be brought to attention that most of the elements are of larger size than the loads would suggest. With such precision requirement the structure cannot cause any significant deformations; therefore, it has to be over-constrained (Fig. 3).

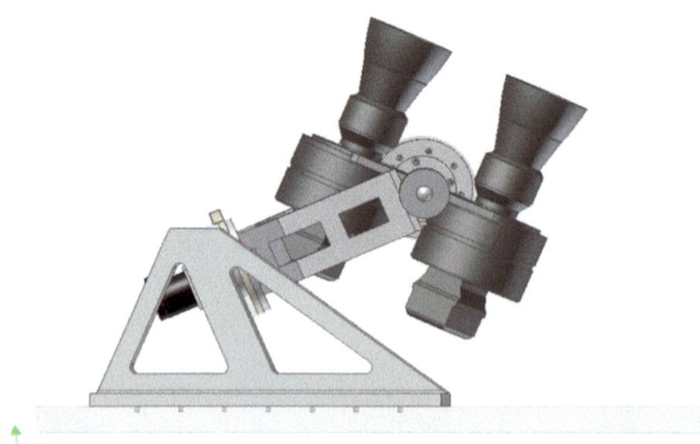

Fig. 5 The case number 1

Fig. 6 The case number 2

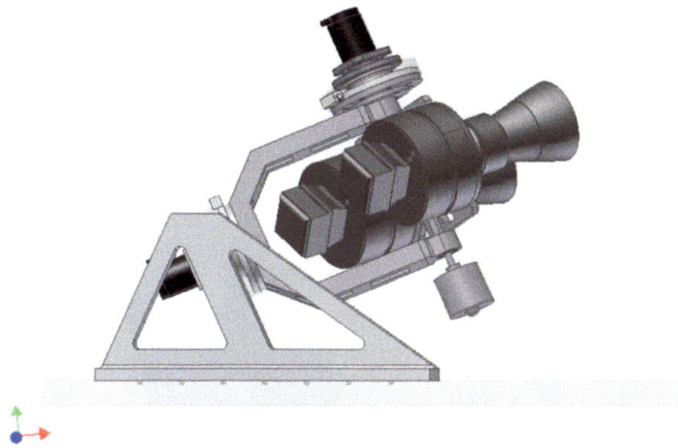

2.4 Analysis of Loads

The structural analysis of the fork (Fig. 4), holding the cameras, is described below. Loads applied to the fork are forces and moments resulting from the weight of cameras, motor, gear and other elements (about 22 kg). Chosen model assumes equal distribution of load on each of the four mounting points. The fork is buttressed in four points, which correspond to mounting points and to top base through an additional fixing plate.

Fig. 7 The results of calculation; case number 1

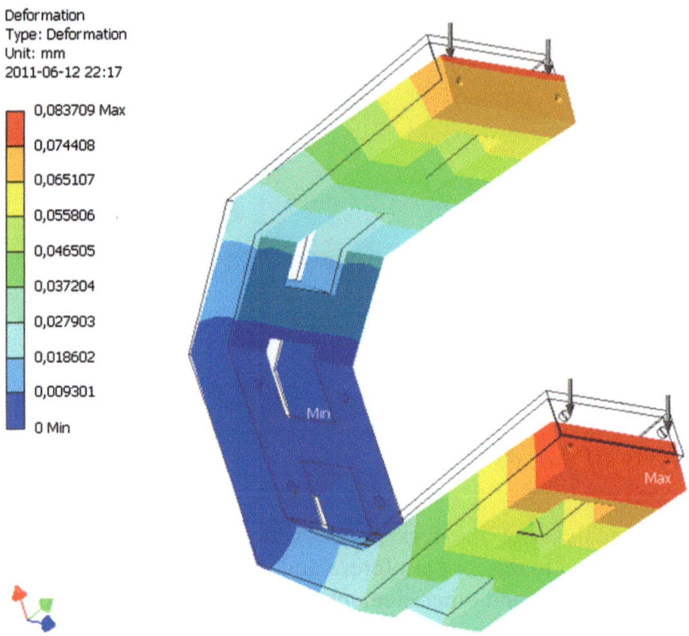

Fig. 8 The results of calculation; case number 2

A cubic mesh (linear discretization) has been used for FEM analyses. The fork was divided into 20,659 elements (Young's modulus E = 210 GPa; Poisson's ratio $v = 0.3$, mass density $\rho = 7.85$ g/cm^3, tensile yield strength 207 Mpa, tensile ultimate strength 345 MPa). Structural analysis was performed for two cases of camera positions. The results are shown below.

Case 1: Arms of the fork have equal loads (Fig. 5)
Case 2: Arms of the fork are positioned "one under another" (Fig. 6)

Analysis Results (Figs. 7, 8):

As shown by the analyses above, the 2nd case results in greater deformations than the 1st one. Results of the analysis allow for calculation of the deformation angle of the line connecting mid-points of fork's arms, whose estimated values are 0,003680 (0,2208') in the second case, and 0,001690 (0,1008') in the first case. In comparison to errors generated by the motor, coupling and gear (3.14725'), the deformation error is insignificant. In order to minimize this angular deviation one could increase the profile of the rectangular tube, which would increase its rigidity. Another option is to abandon the welded tubular profile concept and use stiffer and heavier bars. However, at this point the obtained value is acceptable (Fig. 8).

3 Conclusions

The above project displays that even with the most basic methods a precise drive for astronomic devices can be designed. Thanks to widely available components of high quality it is possible to create a simple and user-friendly apparatus. This plain design guarantees high reliability. The requirements defined at the beginning of the project have been fulfilled—desired accuracy, precision and smoothness of both type of motion, used for constant observation as well as the quick motion for changing of the looking direction, have been achieved.

The project was written as a bachelor's degree thesis. Of course, it is only the first trial and requires more detailed analyses, scripting of proper software and developing the controls. Nevertheless, it can already be said that this type of drive would allow for observation of the sky which would surely satisfy more than one astronomer.

Trajectory Planning and Simulations of the Manipulator Mounted on a Free-Floating Satellite

Tomasz Rybus and Karol Seweryn

Abstract This chapter focuses on the dynamics of a 6 DOF manipulator mounted on a satellite during an orbital servicing mission. Algorithm for computation of control torques applied in manipulator's joints in various parts of orbital rendezvous maneuver is presented. This algorithm can be used to calculate manipulator's control torques along with torques and forces for control of satellite's orientation and position during complex maneuver or for controlling only the manipulator while taking into account additional 6 DOFs of the free-floating satellite. Simulation tool used for performing numerical simulations of orbital rendezvous and results of simulations of two parts of the possible maneuver (capturing of tumbling target satellite and positioning captured satellite for docking) are also presented.

1 Introduction

The lifetime of a considerable number of satellites is significantly shortened by malfunctions occurring during operational period, e.g., failures of attitude control systems, failures of deployment mechanisms for communication antennas and solar arrays, or premature depletion of fuel (Ellery et al. 2008). In recent years, various studies have indicated that technology developments in robotics, automation and control allow performing on-orbit satellite servicing that might prove economically justified (Cougnet et al. 2006, Kreisel 2002). Most of the presented mission concepts rely on the use of a robotic arm for the purpose of capturing a target satellite (e.g., Yasaka and Ashford 1996). Moreover, servicing satellite capable of capturing non-cooperative objects (such as satellites with non-operational attitude control) could also be used for capturing defunct satellites in

T. Rybus (✉) · K. Seweryn
Space Research Centre of the Polish Academy of Sciences, ul. Bartycka 18a 00-716 Warsaw, Poland
e-mail: trybus@cbk.waw.pl

J. Sąsiadek (ed.), *Aerospace Robotics*, GeoPlanet: Earth and Planetary Sciences,
DOI: 10.1007/978-3-642-34020-8_6, © Springer-Verlag Berlin Heidelberg 2013

order to remove them from orbit (Rebele et al. 2002). Removal of large space debris may become a necessity since studies show that current debris population in LEO is likely to increase also due to random collisions between existing on-orbit debris (Liou and Johnson 2006). Even if all of the new satellites were launched with some built-in disposal system, active removal of existing, intact objects (several per year) might be required to prevent growth of the existing LEO population of space debris (Lewis et al. 2009). One should also keep in mind that unplanned and uncontrolled atmospheric re-entry of especially large objects might pose a threat to population on Earth. Furthermore, it is important to note that an application of robotics in various orbital operations may be essential in future development of Space Situational Awareness (Brauer 2006).

The development of a satellite capable of performing autonomous servicing mission is a complex, multidisciplinary and demanding task. Such missions require utilization of many sophisticated technologies, e.g., optical system for identification of target's parameters of motion, and reliable lightweight manipulator capable of capturing objects which are not equipped with dedicated docking ports (Waltz 1993). Specific required technologies were tested with numerous space missions (e.g., ATV) but also some dedicated technology demonstration missions were successfully performed [e.g., ETS-VII in 1997 (Kasai et al. 1999) and Orbital Express in 2007 (Ogilvie et al. 2008)] and other are planned [e.g., Deutsche Orbitale Servicing Mission—DEOS (Rank et al. 2011; Reintsema et al. 2011)].

As the control of the satellite-manipulator system is one of the major challenges in on-orbit servicing, numerical simulations of dynamics of such systems are extremely important. During orbital rendezvous maneuver, motions of the manipulator induce reaction torques and forces acting on its base. These torques and forces affect attitude and position of the servicer satellite. Simulations are, therefore, needed for assessment of these influences during design of the satellite's Guidance, Navigation and Control subsystem (GNC). Numerical simulations are also essential for testing control algorithms, which should take into account free-floating nature of a manipulator-equipped satellite. Conceptual illustration of on-orbit servicing is presented in (Fig. 1).

Fig. 1 Conceptual illustration of on-orbit servicing

The chapter is divided into five sections (including the introduction). In the Sect. 2, trajectory planning algorithm for the manipulator mounted on a satellite is presented. Simulation tool used for performing simulations of orbital rendezvous maneuver is described in the next section. Results of numerical simulations are presented in the section four. The Sect. 4 contains conclusions and plans for future work.

2 Trajectory Planning

In the analyzed scenario of technology demonstration of orbital servicing mission, a 6 DOF manipulator is mounted on a satellite. Three general cases for the considered control algorithm can be distinguished: (1) controlling manipulator and the satellite on which the manipulator is mounted (computation of manipulator's control torques along with torques and forces for control of satellite's orientation and position during complex maneuver), (2) controlling only the manipulator while taking into account additional 6DOFs of the free-floating satellite (manipulator's trajectory is realized, whereas satellite is free to change orientation and position according to reaction torques and forces induced by the manipulator), and (3) controlling fixed-base manipulator (this corresponds to the Earth-case or manipulator mounted on a satellite with active and independent attitude and position keeping system). Evidently, one could also consider mixed cases, where, for example, only orientation of the satellite is controlled during manipulator's maneuver (actually, such case is common).

In the most general case (1), state vector of the satellite-manipulator system consists of components describing satellite position (x_s, y_s and z_s), satellite orientation (Euler angles ϕ_s, θ_s and ψ_s) and positions of manipulator's joints (θ_1–θ_6):

$$\mathbf{q} = [x_s, y_s, z_s, \phi_s, \theta_s, \psi_s, \theta_1, \theta_2, \theta_3, \theta_4, \theta_5, \theta_6]^T \tag{1}$$

In this case we assume that the satellite's GNC is able to control position and orientation of the satellite and inputs for this system are computed along with the manipulator's control torques, hence one algorithm can be used for planning complex satellite-manipulator maneuver. Control vector of the considered system consists of 12 components, i.e., forces (F_x, F_y and F_z) and torques (T_x, T_y and T_z) generated by the GNC and control torques in the manipulator's joints (T_1–T_6):

$$\mathbf{Q} = \left[F_x, F_y, F_z, T_x, T_y, T_z, T_1, T_2, T_3, T_4, T_5, T_6\right]^T \tag{2}$$

Generalized equations of motion (3) for a system consisting of a satellite equipped with 6 DOF manipulator can be derived from the second kind Lagrangian equations (Seweryn and Banaszkiewicz 2008):

$$\mathbf{Q} = \mathbf{M}(\mathbf{q})\ddot{\mathbf{q}} + \mathbf{C}(\dot{\mathbf{q}}, \mathbf{q})\dot{\mathbf{q}} \tag{3}$$

where **M** denotes mass matrix and **C** denotes Coriolis Matrix (orbiting system is in the state of free fall; therefore, no potential forces are included in this equation). The approach presented in this chapter is based on the General Jacobian Matrix (GJM) approach introduced by Umetani and Yoshida (1989). GJM can be expressed as:

$$\mathbf{M}(\mathbf{q}) = \begin{bmatrix} \mathbf{A} & \mathbf{B} & \mathbf{D} \\ \mathbf{B}^T & \mathbf{E} & \mathbf{F} \\ \mathbf{D}^T & \mathbf{F}^T & \mathbf{N} \end{bmatrix} \tag{4}$$

where the submatrices are defined as follows:

$$\mathbf{A} = \left(m_s + \sum_{i=1}^{n} m_i \right) \mathbf{I} \tag{5}$$

$$\mathbf{B} = \left(m_s + \sum_{i=1}^{n} m_i \right) \tilde{\mathbf{r}}_{s\_g} \tag{6}$$

$$\mathbf{D} = \sum_{i=1}^{n} m_i \mathbf{J}_{Ti} \tag{7}$$

$$\mathbf{E} = \mathbf{I}_s + \sum_{i=1}^{n} \left(\mathbf{I}_i + m_i \tilde{\mathbf{r}}_{i\_s}^T \tilde{\mathbf{r}}_{i\_s} \right) \tag{8}$$

$$\mathbf{F} = \sum_{i=1}^{n} \left(\mathbf{I}_i \mathbf{J}_{Ri} + m_i \tilde{\mathbf{r}}_{i\_s} \mathbf{J}_{Ti} \right) \tag{9}$$

$$\mathbf{N} = \sum_{i=1}^{n} \left(\mathbf{J}_{Ri}^T \mathbf{I}_i \mathbf{J}_{Ri} + m_i \mathbf{J}_{Ti}^T \mathbf{J}_{Ti} \right) \tag{10}$$

and

$$\mathbf{r}_{s\_g} = \mathbf{r}_s - \mathbf{r}_g \tag{11}$$

$$\mathbf{r}_{i\_s} = \mathbf{r}_i - \mathbf{r}_s \tag{12}$$

In the above equations, $\mathbf{r}_s$ is the position of servicer satellite center of gravity (CG), $\mathbf{r}_g$ is the position of the manipulator mounting point with respect to servicer satellite CG, $\mathbf{r}_i$ is the position of i-th kinematic pair of the manipulator, m_s is the mass of the servicer satellite, m_i is the mass of i-th manipulator's link, **I** is the identity matrix, $\mathbf{I}_i$ is the inertia matrix of i-th manipulator link, $\mathbf{I}_s$ is the inertia matrix of the servicer satellite, $\mathbf{J}_{Ti}$ is the translational component of manipulator' Jacobian [expressed in inertial Coordinate System (CS)], while $\mathbf{J}_{Ri}$ is the rotational component of this Jacobian.

For the second case, in which we are only interested in proper trajectory following of the manipulator (the servicer satellite is uncontrolled and free to change its position and orientation), we use shorter state vector:

$$\mathbf{q}_m = [\theta_1, \theta_2, \theta_3, \theta_4, \theta_5, \theta_6]^T \tag{13}$$

and shorter control vector:

$$\mathbf{Q}_m = [T_1, T_2, T_3, T_4, T_5, T_6]^T \tag{14}$$

In this case, the mass matrix is defined as:

$$\mathbf{M}_m(\mathbf{q}_m) = \mathbf{N} - \begin{bmatrix} \mathbf{C} \\ \mathbf{F} \end{bmatrix}^T \begin{bmatrix} \mathbf{A} & \mathbf{B} \\ \mathbf{B}^T & \mathbf{E} \end{bmatrix}^{-1} \begin{bmatrix} \mathbf{C} \\ \mathbf{F} \end{bmatrix} \tag{15}$$

The last and the simplest of considered cases is the commonly encountered Earth-case—the manipulator mounted on a fixed base. In this situation, the state vector and control vector are the same as in the previous case, i.e., Eqs. 13 and 14, respectively. The mass matrix is defined as:

$$\mathbf{M}_f(\mathbf{q}_f) = \mathbf{N} \tag{16}$$

Though simple, this case applies for manipulator-independent satellite attitude and position keeping system. This system counteracts reaction torque and forces induced by the motions of the manipulator and for such case the manipulator trajectory should be planned as for fixed-base mounting. The possibility of trajectory planning with Eq. 16 is also quite interesting due to its simplicity which could make it easier to implement for control system on-board satellite.

3 Simulation Tool

The modeling and simulation of space robots is a challenging task as the interactions between separate components (GNC, joint controller, manipulator) of the system must be taken into account. Object-oriented modelling is especially suitable for simulations of space robots, and environments like SYMOFROS, DCAP, ADAMS or DYMOLA are often used (Ferretti et al. 2004). The simulation tool presented in this chapter and used for verification of control algorithm described in Sect. 2 was developed in the frame of a joint project between Space Research Centre of the Polish Academy of Sciences and Astrium ST. This tool is based on a model built in SimMechanics (software based on the Simscape, the platform product for the MATLAB Simulink). The utilization of this platform allowed us to incorporate in one environment all the elements essential in numerical simulations of satellite-manipulator dynamics. Simscape solver works autonomously and creates the physical network from the block model of the system. Subsequently the system of equations for the model is constructed and solved (Haug 1989; Wood and Kennedy 2003).

In the presented simulation tool, servicer satellite and target satellite are modeled as rigid bodies. Specific use of SimMechanics built-in block of 6DOF joint allows for simulation of rotational and translational motion of satellites (i.e., orbital motion of both satellites around the Earth and tumbling motion of uncontrolled target satellite). The overview of Simulink model of the servicer satellite is presented in Fig. 2, while detailed model of one of the manipulator's joints is presented in Fig. 3. Simulink model of the GNC subsystem responsible for keeping attitude is attached to each satellite and SimMechanics joint actuators are used to apply control torques. The robotic arm mounted on the servicer satellite is modeled as a chain of rigid bodies connected by rotational joints. The geometrical and inertial properties of manipulator's links are obtained from CAD model of the proposed robotic arm. The friction forces may play an important role in the behavior of the considered system (Landzettel et al. 2006); therefore, Coulomb and viscous friction forces are modeled in joints. The backlash effect is also taken into account. Various sensors are modeled using predefined Simulink blocks and disturbances can be added in different parts of the system. A joint controller based on linear-quadratic regulation (LQR) is also incorporated in this tool (Seweryn et al. 2011), but was not used in simulations presented in this chapter.

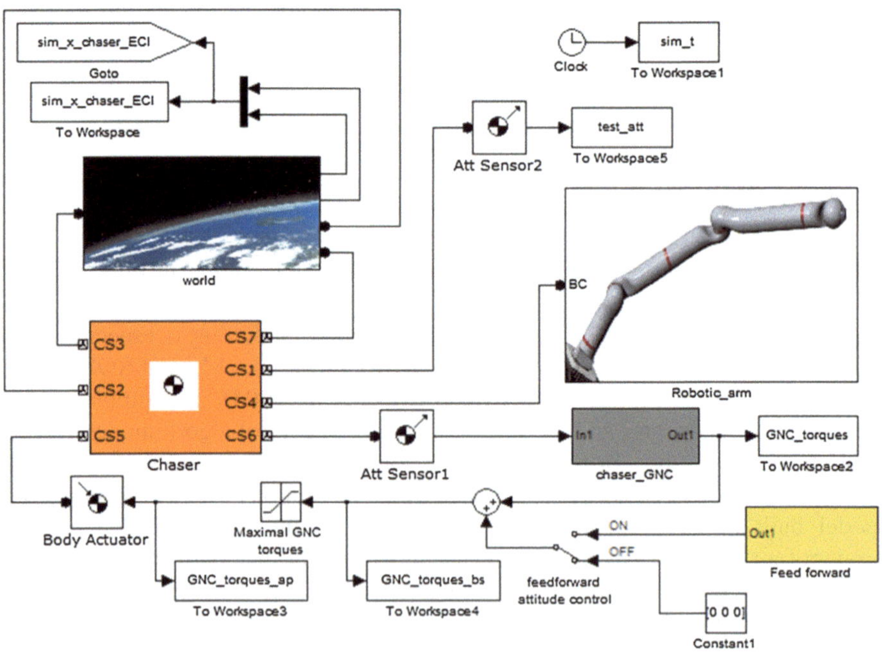

Fig. 2 Overview of Simulink model of the servicer satellite

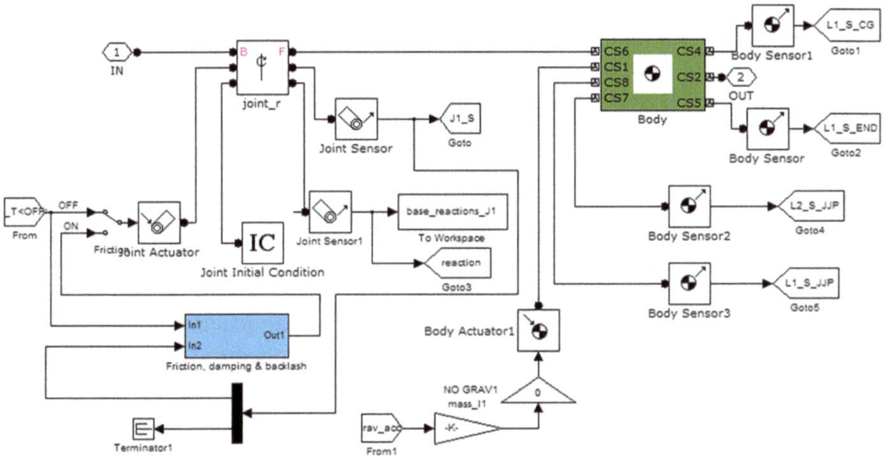

Fig. 3 Simulink model of manipulator's joint

Two stages can be distinguished in the operation of the described simulation tool. During the first stage, a trajectory of the manipulator is planned by the algorithm described in Sect. 2 and implemented in Matlab function. After the phase of trajectory planning, the second stage begins and simulation is performed in Simulink. This serves as a natural verification of trajectory planning algorithms as it is only necessary to check whether or not the trajectory realized by the SimMechanics model of the system is consistent with the planned one.

4 Simulation Results

In the considered scenario of technology demonstration of on-orbit servicing, mass of the servicer satellite is 900 kg. Mass of the captured satellite is 650 kg. The servicer satellite is equipped with 2.96 meter long manipulator that has a mass of 35 kg. Properties of the manipulator's links are presented in Table 1.

Simulation tool described in Sect. 3 allows performing easily simulations with various parameters for different parts of the final rendezvous maneuver. In this

Table 1 Mass and length of manipulator's links

Link no.	Mass [kg]	Length [m]
1	2.36	0.2
2	16.55	1.4
3	13.00	1.1
4	3.35	0.19
5	0.54	0.046
6	0.25	0.021

section we present (as an illustration of torque-computation algorithm described in Sect. 2) exemplary results of simulations of two core parts of such maneuver: (1) capturing tumbling target satellite, and (2) positioning captured satellite for docking with the servicer. The first part of the simulations was performed with independent attitude and position keeping system working on the servicer satellite and, therefore, the third and the simplest version of the presented torque-computation algorithm was used (fixed-base case). In the second part of simulations, this system was simulated as inactive and servicer satellite was free to change its orientation and positions due to the motions of the manipulator. The positioning of a captured object (e.g., in order to attach it to the servicing satellite) can be performed with very low velocity. However, the resulting reaction torques may be substantial, because the mass of the target satellite attached to the end-effector of the manipulator is significant in comparison to the mass of the servicer satellite. As a result, attitude control system might be unable to counteract such high reaction torques. Therefore, in order not to exceed limits of the attitude control, it is better to perform this part of the manoeuvre without active control of satellite's orientation. In this case, the second version of the algorithm was used and additional 6 DOFs of the satellite were taken into account during trajectory planning.

In Fig. 4 orientation (Euler angles in Z-Y-X convention) and angular velocity of the target satellite is presented. As it can be seen in the right panel of this figure, the values of angular velocity components are changing over time due to tumbling nature of motion. The distance between target satellite centre of gravity (CG) and servicer CG is 2.7 m. The distance between manipulator's end-effector and grasping point located on the target satellite is presented in the left panel of Fig. 5. During the course of the manoeuvre, the end-effector is approaching the grasping point in a straight line trajectory (in grasping point CS) and, therefore, only one component of the relative velocity between the grasping point and the end-effector has non zero value. The positions of the first three manipulator's joints that are realizing this motion are presented in the right panel of Fig. 5, while joint control torques computed with presented algorithm are shown in the left panel of Fig. 6. The reaction torques (in respect to servicing satellite CG) induced by the motions of the manipulator during capture manoeuvre are shown in the right panel of Fig. 6. These torques are not causing changes of servicer attitude due to the actions of the attitude control system. The frames from animation presenting this entire part of orbital rendezvous are shown in Fig. 7.

In the left panel of Fig. 8, positions of the first three manipulator's joints are presented for the second part of the considered manoeuvre—positioning captured target satellite for docking with the servicer. As was described earlier, the joint control torques (presented in the right panel of Fig. 8) were computed for the free-floating servicing satellite. The reaction torques during repositioning of the captured target are presented in the left panel of Fig. 9. These torques are not counteracted by the attitude control system and, therefore, they are causing changes in the servicer satellite orientation (the right panel of Fig. 9). The frames from animation presenting positioning of the captured taget satellite are shown in Fig. 10.

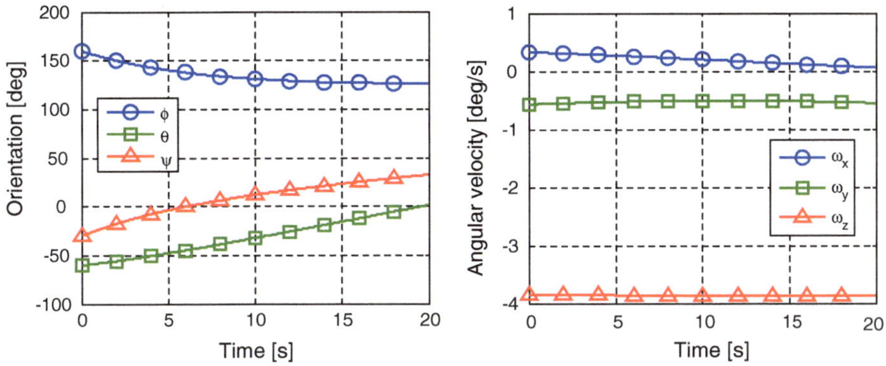

Fig. 4 Target satellite orientation (*left panel*) and angular velocity (*right panel*)

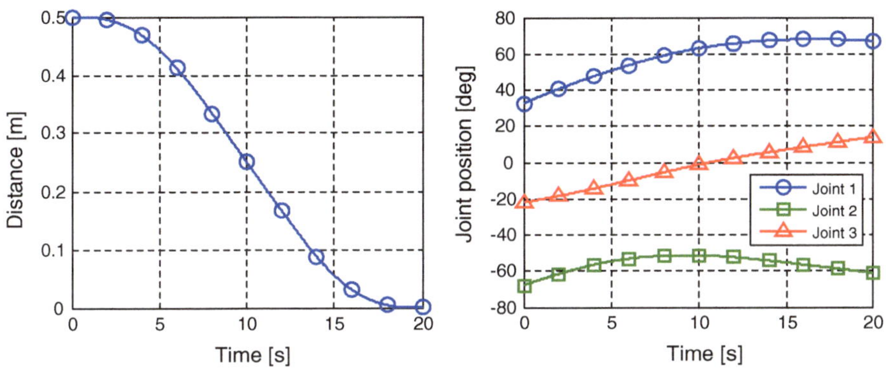

Fig. 5 Distance between the end effector and target satellite docking port (*left panel*) and positions of first three manipulator's joints (*right panel*)

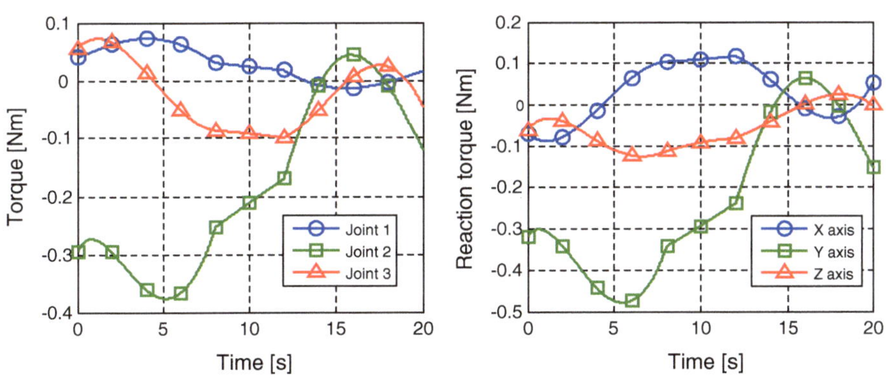

Fig. 6 Control torques applied at the first three manipulator's joints (*left panel*) and reaction torques induced by the manipulator in respect to servicing satellite CG (*right panel*)

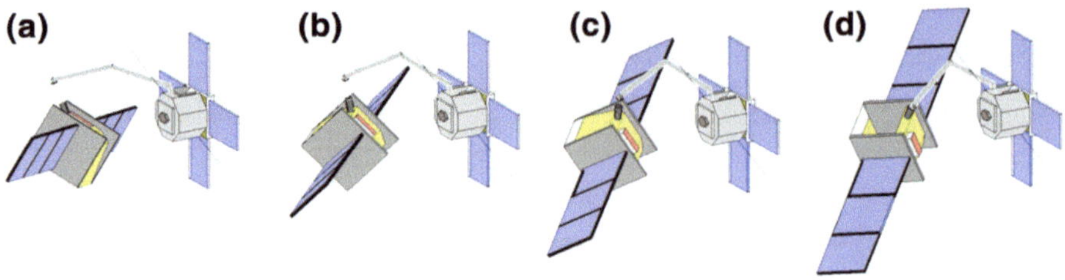

Fig. 7 Frames from the animation presenting the capture of the tumbling target satellite

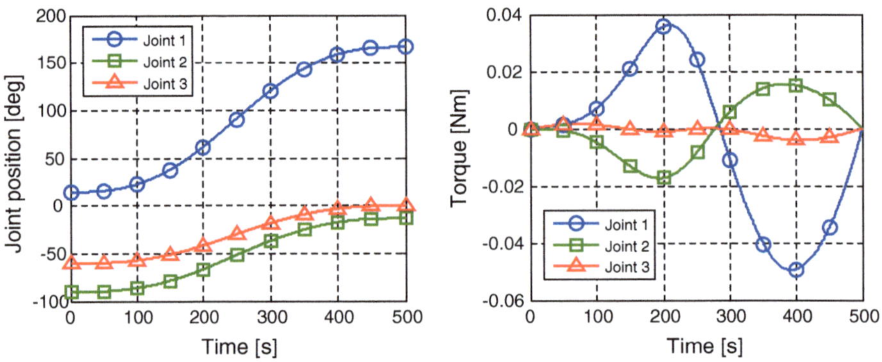

Fig. 8 Positions of the first three manipulator's joints (*left panel*) and the control torques applied at these joints (*right panel*)

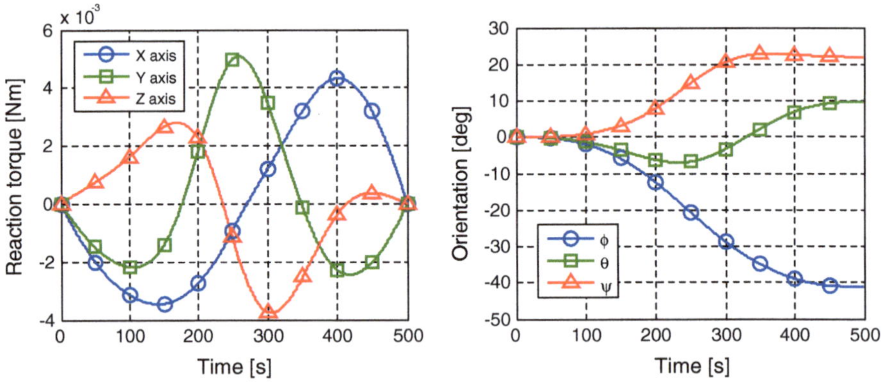

Fig. 9 Reaction torques induced by the motions of the manipulator in respect to the servicing satellite CG (*left panel*) and servicing satellite orientation (*right panel*)

When GNC is active, the attitude control torques are counteracting the reaction torques and satellite is keeping constant orientation (this case was presented in the first part of simulation results). In such case, because of limited freedom of

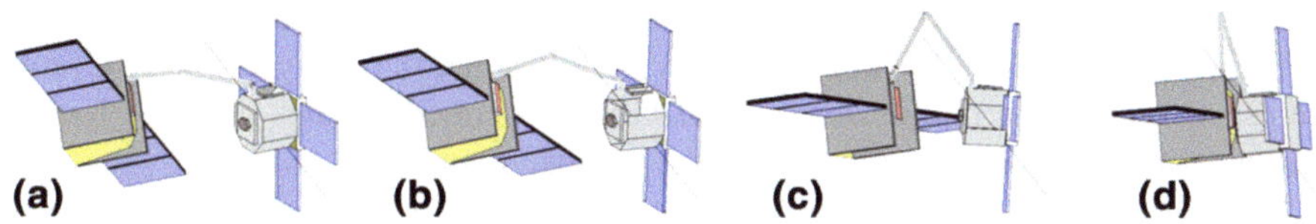

Fig. 10 Frames from the animation presenting positioning of the target satellite for docking with the servicer satellite

satellite, the torques applied at the joints must be higher and, as a result, reaction torques induced by the manipulator are also higher. This shows that the interactions between the manipulator and the attitude control system must be taken into account to obtain reliable results (Fig. 10).

It is also important to note that capturing tumbling target had to be conducted quickly (in simulated scenario this part of the manoeuvre lasts 20 s) in order to catch the grasping point before it moves out of manipulator's range or before the end-effector trajectory will be obstructed by, e.g., target's solar panels. The second part of the manoeuvre, i.e., positioning of the captured target, could last much longer (e.g., 500 s), but heavy load attached to the end-effector would, nevertheless, cause high reaction torques if this part of the manoeuvre were conducted with active attitude control.

The changes in the servicer satellite orientation during manoeuvres, in which free rotations of this satellite are allowed, may result in positioning of the communication antennas out of the required cone. If the temporary loss of communication cannot be afforded, one must make sure that the changes in the servicer satellite orientation are not significant enough to cause such loss. The performed simulations allow us to analyse the influence of the manipulator's motion on the servicer satellite orientation.

5 Conclusions and Future Work

The torque-computation algorithm presented in this chapter (based on the General Jacobian Matrix approach) can be used to calculate manipulator's control torques along with torques and forces for control of the satellite's orientation and position during complex maneuver of the orbital rendezvous. This algorithm can also be used for controlling only the manipulator while taking into account additional 6 DOFs of the free-floating satellite. A flexible simulation tool that takes into account interactions between separate components of servicer satellite (GNC, joint controller, manipulator) created in Simulink environment was used to verify this algorithm. The developed simulation tool and torque-computation algorithm can be used to perform analyses of various parts of the final stage of orbital rendezvous (unfolding of the manipulator, capturing tumbling target, folding of the manipulator with attached target).

One of the purposes of the presented simulation tool is the assessment of the performance of satellite's Guidance, Navigation and Control system (GNC) and the estimation of the maximal values of GNC-generated torques required for the attitude control of the satellite during the capture maneuver.

Regardless of effort that is put into theoretical verification of numerical simulation results, experimental validation can never be completely omitted. It appears that test-bed systems are logical step between numerical simulations and technology demonstration missions. Currently work is in progress at the Space Research Centre of the Polish Academy of Sciences to build a test-bed that would allow testing manipulators in conditions similar to microgravity. This test-bed will incorporate flat round air-bearings based on porous media technology to provide almost frictionless (friction coefficient around 10^{-5}) planar motion. The data obtained from vision system and encoders during an experiment will be compared with values derived from simulations performed by the simulation tool described in this chapter. Although limited to two dimensions, experiments performed with such test-bed could be used to verify simulations and test the proposed control algorithms.

Acknowledgments The simulation tool was developed during joint project of SRC PAS and ASTRIUM ST.

References

Brauer G (2006). European space situational awareness. In: Building the architecture for sustainable space security: conference report. UNIDIR, pp 145–151

Cougnet C, Gerber B, Heemskerk C, Kapellos K, Visentin G. (2006). On-orbit servicing system of a GEO satellite fleet. In: Proceedings of the 9th ESA workshop on advanced space technologies for robotics and automation (ASTRA 2006), ESTEC, Noordwijk, The Netherlands

Ellery A, Kreisel J, Sommer B (2008) The case for robotic on-orbit servicing of spacecraft: spacecraft reliability is a myth. Acta Astronautica 63:632–648

Ferretti G, Magnani G, Rocco P, Vigano L, Gritti M, Rusconi A (2004) Object-oriented modeling of a space robotic manipulator. In: 8th ESA workshop on advanced space technologies for robotics and automation (ASTRA 2004), ESTEC, Noordwijk, The Netherlands

Haug E (1989) Computer aided kinematics and dynamics of mechanical systems, Vol 1, basic methods, Allyn and Bacon, London

Kasai T, Oda M, Suzuki T (1999) Results of the ETS-7 mission—rendezvous docking and space robotics experiments. In: Proceedings of the 5th international symposium on artificial intelligence, robotics and automation in space (SAIRAS), ESTEC, Noordwijk, The Netherlands

Kreisel J (2002) On-Orbit servicing of satellites (OOS): its potential market & impact. In: Proceedings of the 7th ESA workshop on advanced space technologies for robotics and automation (ASTRA 2002), ESTEC, Noordwijk, The Netherlands

Landzettel K, Albu-Schäffer A, Preusche C, Reintsema D, Rebele B, Hirzinger G (2006). Robotic on-orbit servicing—DLR's experience and perspective. In: Proceedings of the 2006 IEEE/RSJ, international conference on intelligent robots and systems (IROS). Beijing, China

Lewis H, Swinerd G, Newland R, Saunders A (2009) Active removal study for on-orbit debris using DAMAGE. In: Proceedings of the 5th European conference on space debris, Darmstadt, Germany

Liou JC, Johnson NL (2006) Risks in space from orbiting debris. Science 311:340–341

Ogilvie A, Allport J, Hannah M, Lymer J (2008) Autonomous satellite servicing using the orbital express demonstration manipulator system. In: Proceedings of the 9th international symposium on artificial intelligence, robotics and automation in space (SAIRAS), Los Angeles, USA

Rank P, Muehlbauer Q, Landzettel K (2011) Automation & robotic payload on DEOS. In: Proceedings of the 11th ESA workshop on advanced space technologies for robotics and automation (ASTRA 2011), ESTEC, Noordwijk, The Netherlands

Rebele B, Krenn R, Schäfer B (2002) Grasping strategies and dynamic aspects in satellite capturing by robotic manipulator. In: Proceedings of the 7th ESA workshop on advanced space technologies for robotics and automation (ASTRA 2002), ESTEC, Noordwijk, The Netherlands

Reintsema D, Sommer B, Wolf T, Thaeter J, Radthke A, Naummann W, Rank P, Sommer J (2011) DEOS—the in-flight technology demonstration of german's robotics approach to dispose malfunctioned satellites. In: Proceedings of the 11th ESA workshop on advanced space technologies for robotics and automation (ASTRA 2011), ESTEC, Noordwijk, The Netherlands

Seweryn K, Banaszkiewicz M (2008) Optimization of the trajectory of a general free-flying manipulator during the rendezvous maneuver. In: Proceedings of the AIAA guidance, navigation and control conference and exhibit, Honolulu, Hawaii, USA

Seweryn K, Banaszkiewicz M, Maediger B, Rybus T, Sommer J (2011) Dynamics of space robotic arm during interactions with non-cooperative objects. In: Proceedinds of the 11th ESA workshop on advanced space technologies for robotics and automation (ASTRA 2011), ESTEC, Noordwijk, The Netherlands

Umetani Y, Yoshida K (1989) Resolved motion rate control of space manipulators with generalized jacobian matrix. IEEE Trans Robot Autom 5(3):303–314

Waltz DM (1993) On-orbit servicing of space systems. Kriger Publishing Company, Florida

Wood G, Kennedy D (2003) Simulating mechanical systems in simulink with simmechanics. Technical report, the mathworks, Inc., Natick, USA

Yasaka T, Ashford W (1996) GSV: an approach toward space system servicing. Earth Space Rev 5(2):9–17

Multibody Modelling of a Tracked Robot's Actuation System

Janusz Frączek, Marek Surowiec and Marek Wojtyra

Abstract A simulation model of a mobile robot is presented in the chapter. The robot is equipped with four track systems, wrapped around four movable and independently driven track holders. Driving torques are transmitted to the track systems and track holders via speed reducers. The study is focused on friction effects in gearing, and especially on the self-locking properties. A simplified mathematical model of friction in speed reducers is presented. The model is based on the Coulomb friction law and exploits the analogy between reducers and wedge mechanisms. This friction model is implemented in a general purpose simulation software in which the entire tracked mobile robot is modelled. A multibody model of the complete robot is briefly described. Simulation results obtained for different friction levels, varying from friction absence to friction beyond the self-locking limit, are compared and discussed. The robot motors are also modelled and requirements for electric power in various operating conditions are estimated.

1 Introduction

Designing and assembling of a Small Mobile Robot (SMR) is one of the tasks of the PROTEUS project (www.projektproteus.pl 2012) which deals with integrated mobile system supporting anti-terrorist and anti-crisis actions. A concept of the robot structure is depicted in Fig. 1. The SMR is planned to be an unmanned platform capable of carrying various sets of equipment (cameras, detectors,

J. Frączek · M. Surowiec · M. Wojtyra (✉)
Institute of Aeronautics and Applied Mechanics, Warsaw University of Technology, Nowowiejska 24, 00-665 Warsaw, Poland
e-mail: mwojtyra@meil.pw.edu.pl

J. Frączek
e-mail: jfraczek@meil.pw.edu.pl

J. Sąsiadek (ed.), *Aerospace Robotics*, GeoPlanet: Earth and Planetary Sciences,
DOI: 10.1007/978-3-642-34020-8_7,

manipulator arm etc.) into potentially hostile and hardly accessible environment. The robot is dedicated to terrestrial operations; however, similar vehicles could be used during interplanetary missions.

The design process is supported by multibody modelling and simulation of the SMR. The objectives of modelling are to study dynamic behaviour of the robot and to estimate power consumption as well as forces and torques acting on the robot (especially driving torques required for typical manoeuvres). The virtual model presented here was built prior to the first hardware prototype. The study is focused on quantitative estimation of power consumption, since batteries of an autonomous vehicle should not be recharged too often. This issue is far more important in the case of extraterrestrial operations. Multibody simulation methods used in this study can be easily applied to space systems analysis. Moreover, due to limited possibilities of performing hardware tests, the simulations can prove to be even more useful.

The robot being designed is equipped with four track systems wrapped around four movable and independently driven track holders (Fig. 1). It is propelled by eight electric motors, and the driving torques are transmitted to the tracks and track holders via speed reducers. The transmitted power is partially dissipated due to friction occurring between cooperating elements of gears. The gears in the track holders power transmission systems are self-locking (i.e., irreversible), thus friction effects play extremely important role (Olędzki 1995; Leonesio and Bianchi 2009).

One of the most important objectives of the SMR simulation is to gain quantitative information on power consumption and torque requirements in the presence of friction, especially in the case of self-locking gears. Since a big portion of transmitted power is dissipated, irreversible speed reducers are characterized by relatively small efficiency. However, when motion is ceased and an external load applies a dynamic or static torque to the output gear shaft, no input torque is required to keep the system in a state of rest (Olędzki 1995; Leonesio and Bianchi 2009). Thus, when motor is stopped, no power is consumed (in the case of reversible gears, a stopped motor must develop a non-zero torque to balance external loads).

This article is focused on studying effects of friction forces occurring in the power transmission system. Reduction gears with different friction levels (ranging from frictionless to self-locking ones) are investigated and requirements for driving torques, as well as for electric power, are estimated. The movable track holders actuation and transmission systems are examined.

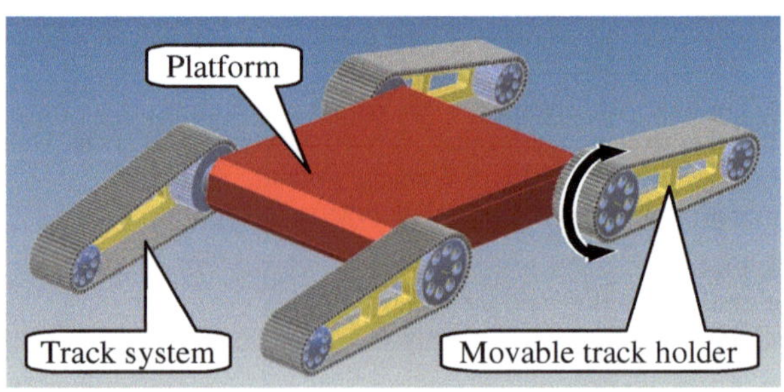

Fig. 1 A small mobile robot (CAD visualisation)

At the early stage of the SMR design process, various types of speed reducers were considered to be installed in the prototype. Therefore, it was decided to develop a model of friction effects in speed reducer which is both quite simple and versatile. The versatility is understood here as ability of applying the same model to simulation of different types of gears, e.g., worm, cylindrical, bevel, epicyclic, etc. It was assumed that the developed model of a speed reducer with friction effects must allow the establishment of appropriate relations between input and output torques, and—obviously—must hold the ratio between input and output angular velocities. Since the model is going to be just a part of a bigger model of the whole vehicle, it is not necessary to delve deeply into construction details of the investigated gearing, and thus building separate models for each type of gearing should be avoided. The mathematical model of friction in speed reducer presented in this chapter is based on the Coulomb friction law and exploits the analogy between a reducer and a wedge mechanism.

In order to account for electric power consumption it was needful to utilise models of DC motors. A radically simplified model of robot control system was also built. The developed models of friction effects within gear train, as well as models of motors and controllers, were implemented in a general purpose simulation software in which the SMR was modelled.

The simulation model of mobile robot actuation and transmission system can operate in two distinct modes. In the first case, driving constraints are imposed on each motor shaft rotation. Due to kinematic nature of motion definition, this mode of model operation is referred to as "kinematic". In the second case, electromechanical properties of motors are taken into account, and driving constraints are replaced by driving torques. This mode of model operation is referred to as "dynamic", and it allows not only testing robot with motors but also considering some control system features.

Simulation results, obtained for different operation modes and different friction levels, are presented and discussed in the chapter. Torque and power requirements during track holders configuration changes are investigated.

2 Mathematical Models of Robot Actuation System Components

2.1 Modelling of Friction in Speed Reducers

It was assumed that friction between cooperating elements of a speed reducer is described by the Coulomb dry friction model. Another assumption is that the gear mechanism is not overconstrained, thus its motion can be uniquely determined using a rigid body model (Frączek and Wojtyra 2011). It was also assumed that masses of moving parts of the reducer are negligible (inertial properties of the reducer shafts can be accounted for by introducing appropriate amendments into inertial properties of vehicle parts connected to the shafts).

In a multibody simulation model an ideal, frictionless gear can be represented by appropriate constraints that enforce proper relations between input and output shafts motion. We decided to use this kind of model and to introduce additional torque responsible for friction effects. Thus, our considerations are focused on friction torque calculation.

The mathematical model of friction in a speed reducer is based on analogy between the reducer and a wedge mechanism. Note that two-sided wedge, depicted in Fig. 2, is introduced in order to allow the normal force to reverse its direction. The analogy between the wedge mechanism and a screw or worm gear is obvious. In the case of other types of gears, the mathematical model based on wedge mechanism equations can be treated as a coarse approximation.

In the simplified model of a speed reducer force **F** corresponds to the driving torque which is applied to the input shaft *I*, and force **P** corresponds to the external load torque which is applied to the output shaft *O*. It is assumed that friction forces are acting only in the kinematic pair formed by bodies *I* and *O*. Since available motor acceleration is limited and masses of gear elements are assumed to be negligibly small, the inertial forces are not taken into account.

The inclination angle α (Fig. 2) is related to the speed reducer gear ratio r (i.e., the ratio between input and output angular velocity) which can be expressed by the following equation:

$$\tan\alpha = 1/r. \tag{1}$$

Let us consider body *I* placed (without clearances) between two inclined planes belonging to body *O*, as shown in Fig. 2. The external force **F** is applied to body *I*, and body *O* acts upon body *I* with normal force **N** and tangent force **T**. Vector

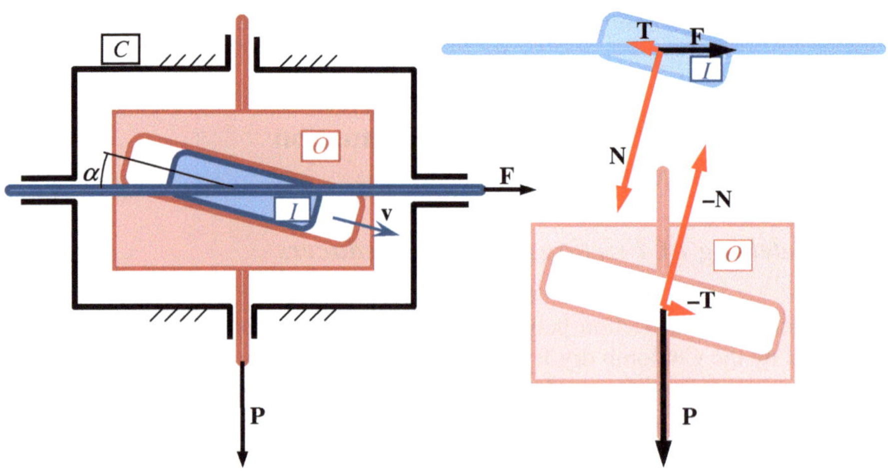

Fig. 2 Distribution of forces in a wedge mechanism

v represents the velocity of body I with respect to O. The external force **P** is applied to body O, and body I acts upon body O with forces –**N** and –**T**. The forces acting between the wedge mechanism casing (body C) and bodies I and O are perpendicular to respective axes of relative motion (and omitted in the figure). Figure 2 presents positive directions of the velocity and force vectors, i.e., if forces are directed as shown in the figure, the corresponding scalar values F, P, N and T are positive, otherwise they are negative (note that, e.g., scalar value F can be treated as a "signed" magnitude of vector **F**, with the sign responsible for indicating the vector direction).

The following scalar equations of force equilibrium can be formulated for body I and O, respectively:

$$F = N \sin \alpha + T \cos \alpha, \tag{2}$$

$$P = N \cos \alpha - T \sin \alpha. \tag{3}$$

Equation (2) multiplied by sin α and added to Eq. (3) multiplied by cos α, yields:

$$N = P \cos \alpha + F \sin \alpha. \tag{4}$$

Let us start with discussing friction from less complicated case of kinetic friction, which occurs when the relative velocity **v** is non-zero. In this case the relation between the normal load and the friction force can be expressed as:

$$T_f = \mu_K |N| \mathrm{sgn}(v). \tag{5}$$

where μ_K is the coefficient of kinetic friction.

The situation is more complicated when the relative velocity **v** equals zero. This is the case of static friction. The maximum possible static friction force can be calculated as:

$$T_{\max} = \mu_S |N|, \tag{6}$$

where μ_S is the coefficient of static friction.

Equation (3) multiplied by sin α and subtracted from Eq. (2) multiplied by cos α, yields the formula for tangent force:

$$T = F \cos \alpha - P \sin \alpha. \tag{7}$$

The external forces applied to the wedge mechanism are balanced by the static friction force, as long as the friction force is less than $T_{\max}$ and greater than $-T_{\max}$. Otherwise, the external forces overcome the force of static friction and cause sliding to occur, thus the static friction transits into kinetic friction (at the moment of transition the relative velocity is zero, however, the relative acceleration is non-zero). Hence, the friction force at zero relative velocity can be calculated using the following expression:

$$T_f = \begin{cases} T_{\max} & \text{when} \quad F\cos\alpha - P\sin\alpha \geq T_{\max} \\ -T_{\max} & \text{when} \quad F\cos\alpha - P\sin\alpha \leq -T_{\max} \\ F\cos\alpha - P\sin\alpha & \text{otherwise} \end{cases} \tag{8}$$

The considerations presented so far are valid for any values of coefficients of friction, thus the equations can be utilised for both self-locking (irreversible) and back-driving (reversible) gears. Since the self-locking is a very important property of power transmission system, the conditions for self-locking of the considered wedge mechanism should be discussed.

The wedge mechanism is a self-locking one when the coefficient of friction is big enough. The dynamic self-locking occurs when for input force equal zero ($F = 0$) the magnitude of kinetic friction force [see Eq. (5)] is greater than the magnitude of tangent component of the external forces [see Eq. (7)]. Two important cases should be considered. The first, when force **P** is directed upwards (thus, in accordance to the discussed earlier sign convention, $P < 0$) and the body O is moving upwards as well (thus, $v > 0$). In the second case, the force is directed downwards ($P > 0$) and the body O is also moving downwards ($v < 0$).

In the first case, the condition for self-locking can be expressed as:

$$\mu_K |N| \operatorname{sgn}(v) > F\cos\alpha - P\sin\alpha. \tag{9}$$

Substituting $F = 0$ and Eq. (4) into Eq. (9), one obtains (for $P < 0$, $v > 0$):

$$-\mu_K P\cos\alpha > -P\sin\alpha, \tag{10}$$

and then (noticing that $-P > 0$):

$$\mu_K > \frac{\sin\alpha}{\cos\alpha} = \tan\alpha. \tag{11}$$

In the second case (note that this time friction force acts in opposite direction, thus inequality sign in Eq. (9) must be reversed), analogous considerations lead to exactly the same self-locking condition (11).

The static self locking occurs when for input force equal zero ($F = 0$) the magnitude of static friction force [see Eq. (8)] is greater than the magnitude of tangent component of the external forces. The condition for static self locking is similar to Eq. (11); this time, however, the coefficient of static friction is taken into account:

$$\mu_S > \tan\alpha. \tag{12}$$

2.2 Modelling of DC Motors

Permanent magnet DC motors are used to actuate the mobile robot. Mechanical parts of the motors, i.e., stators and rotors, are modelled in the multibody software (which automatically formulates and solves equations of motion), whereas equations describing electrodynamics of the motors must be derived and appended to the multibody model.

A simplified DC motor circuit diagram is presented in Fig. 3. Parameters R and L represent resistance and inductance of armature winding, respectively. U_e represents the back-emf (electromotive force) which is generated when the rotor revolves. Thus, the following equation of electric circuit can be formulated (Hughes 2006; Krause 1986):

$$U = U_e + L\,dI/dt + RI, \tag{13}$$

where U is the voltage of power supply, I is the armature current and t denotes time.

The back-emf U_e is proportional to the rotor angular velocity ω and the generated torque M is proportional to the current through the windings, hence:

$$U_e = K_e \cdot \omega, \quad M = K_m \cdot I, \tag{14}$$

where K_e and K_m are the voltage and torque constants, respectively (note that $K_e \approx K_m$).

Equations (13 and 14) can be combined to obtain:

$$dI/dt = (U - K_e\,\omega - RI)/L. \tag{15}$$

The above equation is integrated during direct dynamics calculations (for each motor). Then driving torque M is calculated using Eq. (14). This torque, acting between the stator and the rotor, is transmitted via the speed reducer to the movable track holder.

Instantaneous electric power consumption can be calculated as:

$$P_E = UI. \tag{16}$$

In a special case of halted motor and constant motor load, Eqs. (14 and 15) (for $\dot{I} = 0$ and $\omega = 0$) combined with Eq. (16) result in the following:

$$P_E = RI^2 = \frac{R}{K_m^2} \cdot M^2. \tag{17}$$

Thus, when the motor operates under stall conditions, the electric power is proportional to the square of generated torque.

Instantaneous mechanical power consumption can be calculated as:

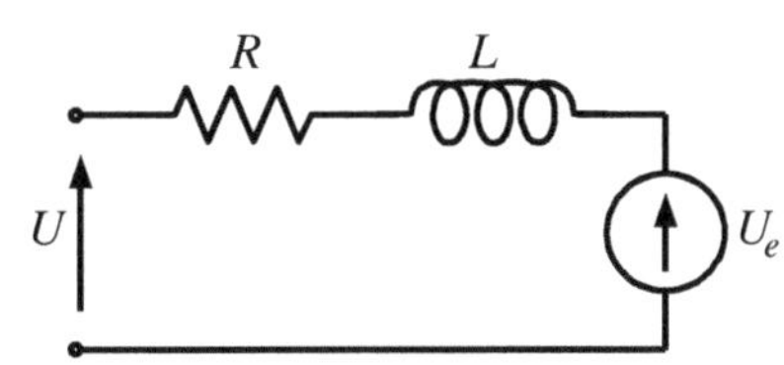

Fig. 3 A simplified circuit diagram of permanent magnet DC motor

$$P_M = M\,\omega. \tag{18}$$

Note that mechanical and electric power requirements can be quite different, especially when the motor develops torque at zero angular velocity.

2.3 Modelling of Controllers

Output torque generated by an electric motor is a function of input voltage, armature current and rotor angular velocity (Hughes 2006, Krause 1986). Robot control system collects information on the current state of the vehicle (supplied by sensors) and compares it with information on the desired state. The control process, resulting in supplying the motor with input voltage necessary for tracking the desired trajectory is usually quite complicated in a real robotic system (Siciliano et al. 2008). In order to focus the study on friction effects in gearing, the model of motors and control system was significantly simplified.

It was assumed that the input voltage U is described by the following equation:

$$U = k_\varphi(\varphi_D - \varphi) + k_\omega(\omega_D - \omega), \tag{19}$$

where φ_D is the desired angle of track holder rotation (with respect to the robot basis), φ is the actual angle of rotation, ω_D and ω are the respective angular velocities, k_φ and k_ω are constant values. Note that Eq. (19) corresponds to PD control algorithm (Ogata 2009).

In the real control system, the pulse-width modulation (PWM) technique is used (Holmes and Lipo 2003); thus, U given by Eq. (19) represents the effective rather than the instantaneous voltage. The effective input voltage may vary between $U_{\min}$ and $U_{\max}$; thus, Eq. (19) must be amended to account for control signal saturation:

$$U = \begin{cases} U_{\max} & \text{for} \quad k_\varphi(\varphi_D - \varphi) + k_\omega(\omega_D - \omega) \geq U_{\max} \\ U_{\min} & \text{for} \quad k_\varphi(\varphi_D - \varphi) + k_\omega(\omega_D - \omega) \leq U_{\min} \\ k_\varphi(\varphi_D - \varphi) + k_\omega(\omega_D - \omega) & \text{otherwise} \end{cases} \tag{20}$$

In the case of irreversible speed reducers, one more modification is introduced to the method of input voltage calculation. It is obvious that control error $(\varphi_D - \varphi)$ is a small value but seldom it is exactly equal zero. Thus, when the motor is stopped and static friction locks the reducer, the input voltage calculated using Eq. (20) is non-zero. Applying this voltage would result in generating torque too small to overcome friction forces, which is pointless (the control error would not decrease). That is why, in the case of irreversible reducers, the method of input voltage calculation is modified: when the motor is stopped and simultaneously the desired angular velocity ω_D is zero, the input voltage, originally given by Eq. (20), gradually decreases its magnitude to zero. When the desired velocity becomes a non-zero value, Eq. (20) is again used in input voltage calculations. Similar procedure is applied in the case of speed reducers with friction below the self-locking limit: the magnitude of input voltage gradually decreases to the minimal value required to keep the reducer in the state of rest. This time, however, this minimal value is greater than zero.

3 Multibody Model of the Small Mobile Robot

3.1 Simulation Model of Tracked Vehicle Dynamics

A commercial multibody simulation package (MSC.ADAMS 2008) was used to build a model of the SMR (Fig. 4). The robot is treated as a system of bodies interconnected by kinematic pairs or contact elements. The modelled bodies are rigid, whereas flexibility effects are considered in auxiliary elements, like springs, bushings, etc. The simulation software allows dynamic analysis of the SMR. Equations of motion for multibody system are formulated automatically; however, equations describing gearing friction, as well as electric motors and controllers must be separately implemented in the software.

A typical multibody model consists of bodies, joints and applied forces. Usually, the kinematic structure of a multibody system does not change. However, when a tracked vehicle is considered, track segments frequently change their connectivity, and kinematic structure of the model changes almost continuously. Each track segment is connected with two neighbouring segments, and—depending on its current position—it can be separated from other bodies, or it can remain in contact with the terrain or with one of the suspension wheels, or with both—wheel and terrain at the same time. Consequently, to avoid kinematic structure changes in the multibody model, interactions between track segments and other bodies are introduced as contact forces rather than as kinematic constraints (Slättengren 2000).

Our present study is not focused on track systems modelling, however—in general—modelling of track contact forces is crucial for simulation reliability and

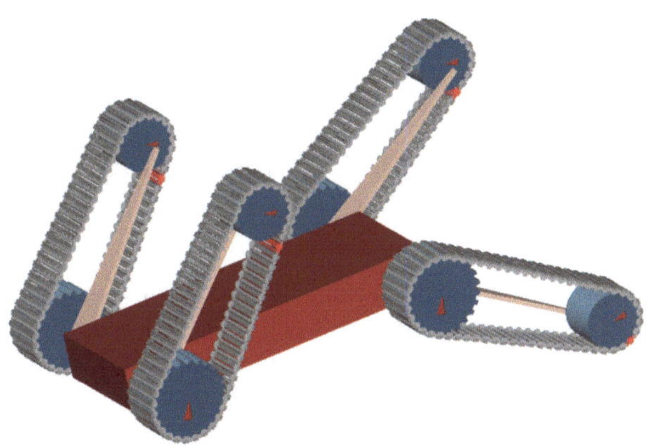

Fig. 4 A multibody model of the SMR

efficiency. The rigid body contact model (MSC.ADAMS 2008) used during simulations is relatively simple, being a reasonable compromise between accuracy and effectiveness. In the vehicle model the number of bodies and contact forces is large, mainly due to large number of track segments, and thus the usage of more sophisticated and mathematically more complex model of contact phenomena would result in very long simulation time.

The simulation software analyzes shapes and instantaneous positions of these parts for which contact is declared, detects contact occurrences and, at the points of contact, determines the penetration depth and velocity. The contact force can be decomposed into two components: normal and tangent to the surface of contact. The normal component is given by the following equation:

$$F = \begin{cases} \max(0, k \cdot x^e - c \cdot \dot{x}) & x > 0 \\ 0 & x \leq 0 \end{cases}, \tag{21}$$

where x denotes the penetration depth. Stiffness k and exponent e are constants specific to the modelled problem [e.g., they depend on terrain and track properties (Wong 2008)]. The damping coefficient c is a function of penetration depth:

$$c(x) = \begin{cases} 0 & x \leq 0 \\ c_{\max}\left(3x^2/h^2 - 2x^3/h^3\right) & 0 < x \leq h \\ c_{\max} & x > h \end{cases}, \tag{22}$$

where the distance h and maximum damping $c_{\max}$ are constants specific to the modelled problem.

The tangent component of contact force is calculated using the Coulomb friction model.

There are no straight rules telling how to choose parameters of the contact model. If it is possible, the simulation results should be—first of all—confronted with experimental data. In our case satisfactory parameters were found after a number of test simulations; the results were compared with data available for similar vehicles. The simulations also showed that the results are not highly sensitive to changes of the values of contact parameters. Thus, even if the parameter estimation is not very accurate, the simulation results reliability is not endangered.

3.2 Simulation Models of Speed Reducers, Motors and Controllers

As it was explained earlier, the model of a generic reducer is focused on establishing proper relations between input and output torque, as well as between input and output speed (and thus, the model reflects power flow and losses). Construction details of the simulated reducers (bearings, gear teeth, etc.) are neglected or modelled in a simplified way. The presented method of modelling can be applied to various types of speed reducers.

From the point of view of a software user, the reducer model consists of three bodies: the input and output shafts and the casing (Fig. 5), regardless the actual type and complexity of the investigated reducer. Masses and moments of inertia of the reducer bodies are negligibly small when compared with inertia of the other parts of simulated vehicle. Moreover, inertial properties of vehicle parts connected to the reducer can be modified to account for the reducer parts inertia. Equations presented in Sect. 2.1 are used to calculate friction torque. In the simulation model this torque is applied to the input shaft as an additional load.

The simulation model is parameterized. The basic parameter is the gear ratio r. Parameter μ is responsible for friction effects modelling; for the sake of simplicity, the same parameter is used for both static and kinetic friction. In the simplified mathematical model of a wedge mechanism, μ is simply the coefficient of friction. In the case of real gears modelling, this parameter may additionally depend on mechanism geometry. Any non-negative value of parameter μ is acceptable in the simulation model, thus it is possible to simulate a wide range of reducers, from frictionless to self-locking ones.

The simulation model can operate in a "kinematic" or "dynamic" mode. In the "kinematic" mode, driving constraints are imposed on angular positions of input shafts (DC motors are not modelled). Note that driving constraints are imposed only on input shafts, and many degrees of freedom are not constrained, thus for the whole vehicle a dynamic (not kinematic) analysis is performed. In the "dynamic" mode driving torques are applied to rotors of DC motors and then transmitted to input shafts. Obviously, a dynamic analysis is performed.

An important difference between two modes of model operation can be observed in the case of static friction occurring in the speed reducer. According to Eq. (8), the static friction torque balances the resultant of input and output torques. In the "dynamic" mode, the static friction torque can be uniquely determined, since the input torque is known. However, when driving constraints are imposed in the "kinematic" mode, the input torque cannot be uniquely determined, since the same motion (namely, the state of rest) can be obtained for various values of the input torque. Thus, in the "kinematic" mode, the static friction torque can have an arbitrary value between the minimum and maximum possible.

The simulation model of a reducer was built in such a way that in the case of static friction (when operating in the "kinematic" mode) among all possible driving torques the one with the smallest magnitude is chosen. This additional

Fig. 5 Simulation model of a speed reducer

condition is very important, since it makes the solution unique. Moreover, this condition is technically justified because the chosen input torque corresponds to the minimum electric power requirements.

It should be emphasized that, from the physical point of view, the problem of finding the input torque, in the case of a state of rest imposed by driving constraints, is not uniquely solvable. Solutions fairly different from those predicted by the simulation model are acceptable as well.

The simulation software automatically generates only the equations describing mechanical properties of the model. Thus, the equations responsible for controllers and electrodynamics of motors were derived (see Sects. 2.2 and 2.3) and then included in the model ("dynamic" mode). At each numerical integration step, input voltage and driving torque are calculated using Eqs. (20 and 14), respectively. The armature current is governed by Eq. (15) which is integrated alongside with other differential equations for the whole multibody model.

4 Simulations

4.1 Objectives and Variants of Simulations

Robot motion along a horizontal plane was investigated during test simulations. The robot was changing configuration of its track holders when moving forward with constant velocity. The motion pattern was the same for both sides of the robot. The simulation study was focused on friction effects in the track holders transmission systems. Simulations for different levels of speed reducers friction, ranging from zero to above the self-locking limit ($\mu = \{0, 0.01, 0.02, 0.03\}$), were performed, whereas the gear ratio ($r = 50$) remained unchanged in all tests. During the first series of simulations, the model was operating in the "kinematic" mode, and during the second series in the "dynamic" mode. Among other quantities, the input torques necessary to fulfil the robot motion requirements were calculated. Mechanical and electric power requirements were also computed.

Note that some parameter values used during simulations were chosen just to illustrate the discussed problems. These parameters do not reflect the actual properties of the designed SMR.

The analysed robot motion lasts ten seconds and can be divided into five phases. During the first phase (0–2 s), front and rear track holders are directed forward. Next (2–4 s), the rear track holders rotate backward. During the third phase (4–6 s) robot configuration remains unchanged. Then (6–8 s), all track holders rotate to move robot platform upwards, and finally, during the fifth phase of motion (8–10 s) the robot configuration is kept unaltered. The sequence of track holders configuration changes is presented in Fig. 6, and the assumed courses of the front and rear track holders angular positions with respect to the robot basis are presented in Fig. 7.

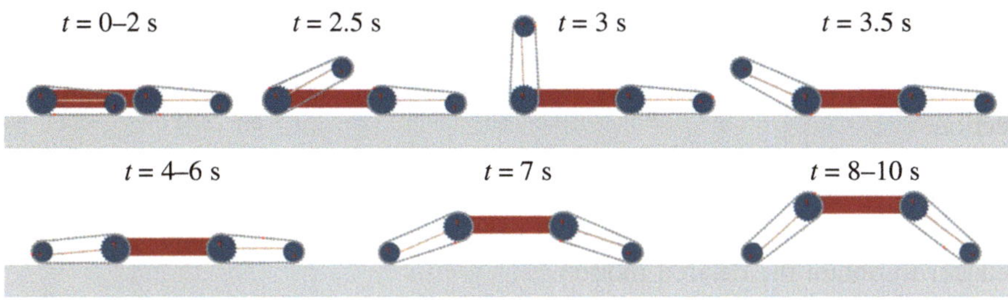

Fig. 6 A sequence of track holders configuration changes

Fig. 7 Desired angular positions of the front and rear track holders versus time

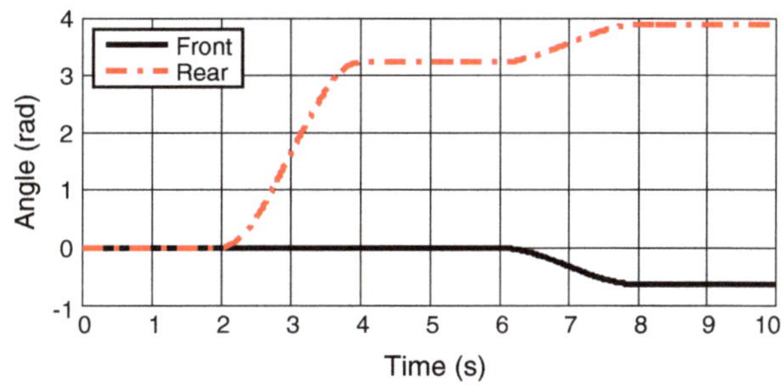

It should be pointed out that in all "dynamic" mode simulations the same desired motion was required; however, in each case the gear friction level was different, thus the controllers acted differently. Therefore, when robot motion is concerned, negligibly small differences between results obtained for subsequent simulations were observed.

4.2 Motion as a Result of Imposing Driving Constraints

During the first series of tests, driving constraints were imposed on the reducers input shafts motion (the required angular positions of track holders are presented in Fig. 7), and the resulting motion of the robot was calculated. Thus, the models of reducers were operating in the "kinematic" mode (recall that term "kinematic" concerns only the reducer input; for the whole system dynamic analyses were performed). The coefficient of friction value was different in each simulation. The driving torques, calculated as reactions of driving constraints, were observed during simulations.

It should be pointed out that, in general, simulations in the "kinematic" mode are intended to be performed prior to selecting motors, in order to support the designer with data necessary to make this selection. Thus, the objective is to estimate requirements for driving torque and mechanical power during typical robot manoeuvres. The requirements for electric power cannot be established in the “kinematic” mode. We decided to concentrate on demanded torque and not to discuss the requirements for mechanical power in this section; both mechanical and electric power requirements are presented and compared in the following section.

The first simulation was performed for frictionless reducers, thus the value of parameter μ was set to zero. Line labelled as “$\mu = 0$” in Fig. 8 presents the driving torque which must be applied to the input shaft of the rear track holder reducer to obtain the desired motion.

As it could be expected, constant values of torque correspond to the phases of robot motion when track holders configuration remains unchanged, and non-constant values of torque are observed when the robot configuration varies. A few important details should be discussed.

Firstly, a rapid change in the input torque value is observed at $t = 2$ s, i.e., when the track holder starts to rotate. At this moment the area of track-ground contact decreases quickly and the centre of pressure moves swiftly backward to find its new position just under the track holder axis of rotation. Similar rapid changes of torque are observed around $t = 4$ s and $t = 6$ s when significant changes of centre of pressure position occur.

Secondly, when the track system is lifted and only the track segments under the driving wheel (the sprocket) are in contact with the ground, as it is observed for $t = 2$–4 s, the moment of ground forces about the track holder axis of rotation is close to zero. Hence, during this phase of motion, the external moment acting on the track holder results mainly from its own weight. That is why, as long as the track holder is moving upwards (2–3 s), the calculated torque is positive, whereas when the track holder is moving downwards (3–4 s), the torque is negative. It is worth noting that for $t = 3$–4 s the direction of input torque is opposite to the direction of rotation; thus, the motor is working as a brake (without the reverse torque, the track holder would accelerate due to gravity forces).

Thirdly, a non-zero torque is required to keep the track holder position unchanged. The observed torques are diverse for different robot configurations, since in each configuration the distribution of mass as well as ground reaction forces is different.

Finally, it is observed that input torque is not constant just after $t = 4$ s despite the fact that the track holders are not moving with respect to the base. To explain

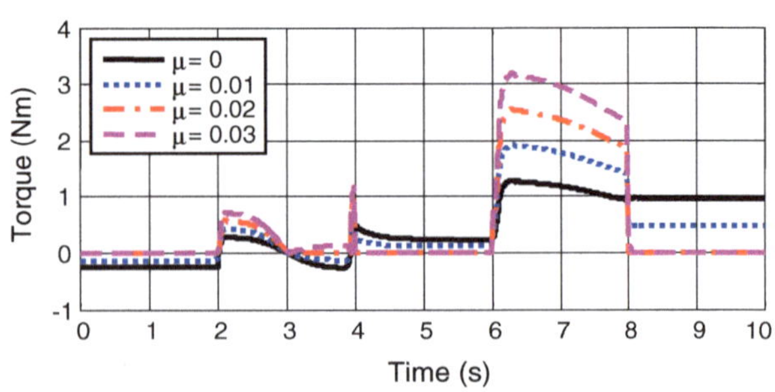

Fig. 8 Input torques for different friction levels (the “kinematic” mode)

this phenomenon we shall recall that track-ground contact is modelled in terms of spring-damper forces [see Eq. (21)]; thus, after configuration changes the system needs some time to find a state of equilibrium. That is why torques exerted on track holders are non-constant for a while just after $t = 4$ s.

The rear track holder speed reducer output torque at every time instant is simply $r = 50$ times larger than the input torque calculated for a frictionless reducer. It should be emphasised that—for the given robot motion—the reducer output torque should be always the same, regardless of the gear ratio and efficiency. Indeed, in our simulations the calculated output torques were almost identical for all tested levels of gear friction (and for both "kinematic" and "dynamic" modes of simulation), since only negligible differences in simulated motion o were observed.

The second simulation was performed for the coefficient of friction $\mu = 0.01$, i.e., below the self-locking limit. The results are presented in Fig. 8 and labelled as "$\mu = 0.01$". Qualitatively the results are similar to those obtained for the case of a frictionless reducer. To analyse effects of friction in gearing, the quantitative differences between lines "$\mu = 0$" and "$\mu = 0.01$" should be discussed.

It can be observed that in the case of reducers with relatively low friction, non-zero torque must be applied when the track holder is not rotating (0–2, 4–6, 8–10 s); however, magnitudes of required torques are much less then in the case of frictionless reducers. Situation is more complicated when the track holder is moving with respect to the base. When the motor lifts the track holder (2–3 s) or the whole robot (6–8 s) the demands for input torque are greater than in the frictionless case, since the motor must overcome friction forces. On the contrary, when the motor serves as a brake (3–4 s), the magnitude of required torque is less than in the frictionless case (when the track holder is moving downwards, the reducer friction forces are supporting the motor).

The third simulation was performed for the coefficient of friction $\mu = 0.02$ which is exactly the self-locking limit, i.e., the reducer operational properties are placed at the border between self-locking and back-driving. The input torque calculated during this simulation is presented in Fig. 8 and labelled as "$\mu = 0.02$". The obtained results are significantly different from results gained in previous simulations. It is noticeable that the torque required to maintain constant position of the track holder is practically equal to zero. Moreover, the input torque required to decelerate the downwards moving track holder (i.e., when the motor is working as a brake) is also very small. On the other hand, torque required to lift the track holder (2–3 s) or the whole robot (6–8 s) is greater than in the previous cases. Similarly, the peak of torque observed around $t = 4$ s (when the rear track holder hits the ground) is greater than the earlier ones.

During the last simulation, the modelled reducer was a self-locking one, with the coefficient of friction $\mu = 0.03$. The calculated input torque is presented in Fig. 8 and labelled as "$\mu = 0.03$". Three important issues should be pointed out in this case. Firstly, even when the track holder is moving downwards (3–4 s), it must be propelled. Unlike in the earlier simulations, during this phase of motion the direction of driving torque and the direction of motor rotation are concordant. Secondly, when the track holder is stopped the required input torque is zero. It means that, since the gearing is irreversible, external loads applied to the stopped track holder are not transmitted back to the motors. Finally, the magnitude of torque required to lift the robot (6–8 s) is greater than in all previous cases. It should be reminded that in the case of static friction (0–2, 4–6, 8–10 s) other input torque solutions are possible. As it was explained in Sect. 3.2, the minimum-magnitude solutions were chosen.

When analysing simulation results compared in Fig. 8, one should notice that taking into account friction forces in gearing is crucial for results credibility, especially when driving torques are solved for. Neglecting friction may result in calculating driving torque not only of a wrong magnitude but also of a wrong direction.

4.3 Motion as a Result of Control and Actuation

When simulation in "dynamic" mode is performed, equations describing motors and controllers (Sects. 2.2 and 2.3) are included in the model. In that case the speed reducer input torque results from control voltage, and not from driving constraints imposition. Apart from driving torques, both mechanical and electric power requirements can be calculated.

Simulation tests similar to described in the previous section were performed. This time, however, the model was operating in the "dynamic" mode. Four simulations, for four different friction levels (same as previously), were carried out. The driving torques obtained from simulations are shown in Fig. 9.

The results presented in Fig. 9 (the "dynamic" mode) are similar to the results shown in Fig. 8 (the "kinematic" mode), and practically all comments presented in the previous section could be repeated here. The only important differences can be observed just after $t = 4$ s and $t = 8$ s. In the case of reducers with non-zero friction, the torques decline to appropriate constant values gradually, and not immediately, as it was observed in the "kinematic" mode. This is the effect of decreasing the input voltage to the minimal required value (see Sect. 2.3). Moreover, the peak of torque observed around $t = 4$ s is smaller, since—due to control process nature—the realised track holder trajectory is similar but not

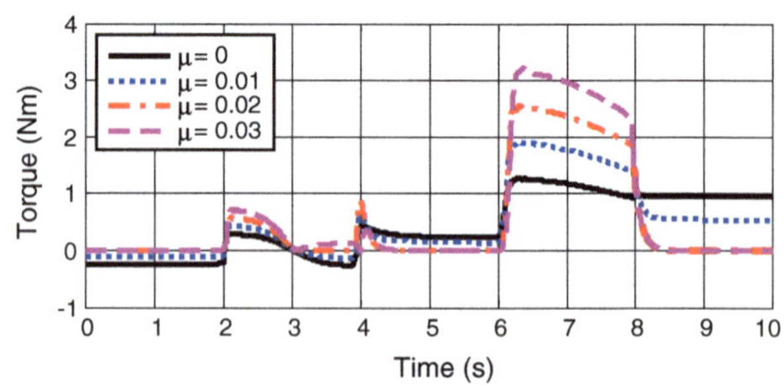

Fig. 9 Input torques for different friction levels (the "dynamic" mode)

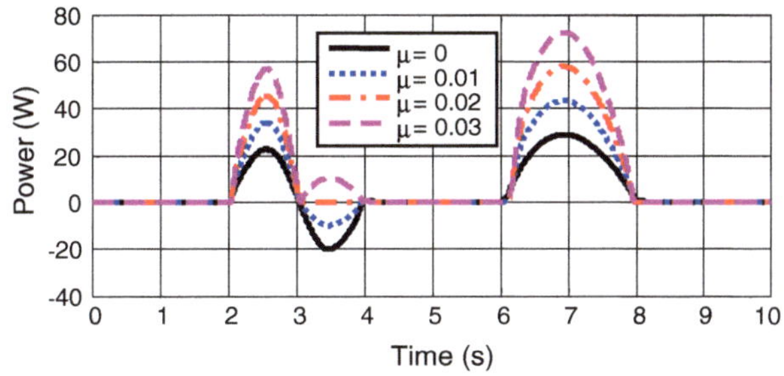

Fig. 10 Mechanical power consumption for different friction levels

identical to the desired trajectory, and thus the changes of track-ground contact area go smoother.

The requirements for mechanical power [see Eq. (18)] for different friction levels are presented in Fig. 10. Obviously, when the track holder is not rotating, the mechanical power equals zero. As it should be expected, the demand for input power increases when friction level increases. The speed reducer output power was in each case the same, regardless of the gear friction. Note that in the case of a frictionless reducer the input and output power are equal, thus the time course of the output power can be read from Fig. 10.

Gear efficiency is usually defined as ratio of output power to input power. The concept of efficiency is useful when the direction of power transmission is constant and when motion without stops is concerned. In the case of robotic applications, however, the idea of efficiency ratio is less useful. Note that proportions between input and output power are completely different, e.g., for $t = 2$–3 s and $t = 3$–4 s. That is why the speed reducer should be analysed in terms of direct comparison of input and output power rather than in terms of gear efficiency.

The requirements for electric power [see Eq. (16)] for different friction levels are presented in Fig. 11. It is worth noting that—in the case of reducers with friction below self-locking level—even when the track holder is not rotating, the electric power is non-zero. As it should be expected, when the track holder is rotating, the demand for input power is higher at higher friction levels.

It should be noted that efficiency of robot drive system depends not only on gear friction, since electric motor operating conditions are also very important.

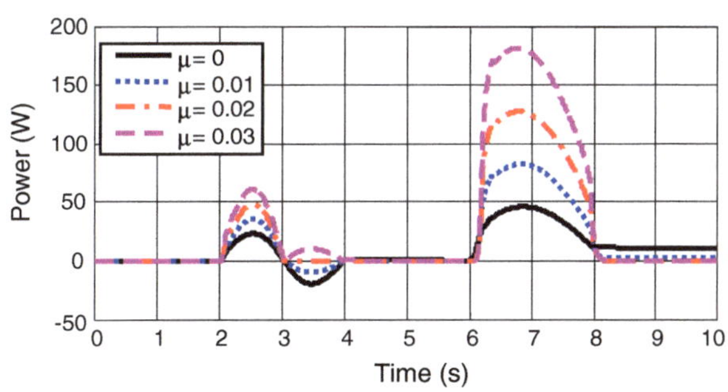

Fig. 11 Electric power consumption for different friction levels

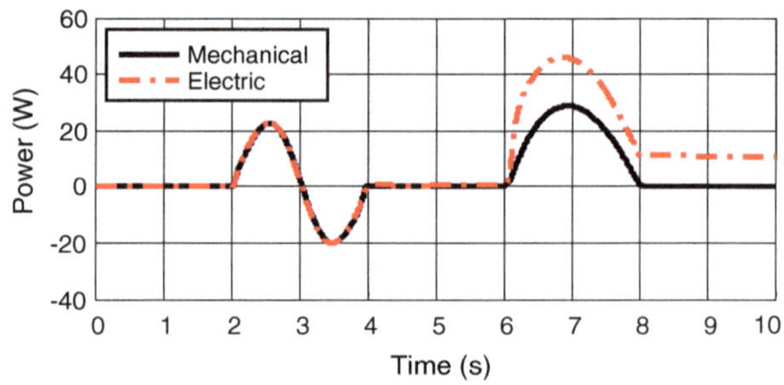

Fig. 12 Mechanical and electric power requirements for a frictionless speed reducer

Figure 12 presents comparison of mechanical and electric power requirements for a frictionless speed reducer. Different types of relations between power demands can be distinguished. When the motor rotates at a high angular velocity and develops a small driving torque (2–4 s), the electric power is close to the mechanical power. On the contrary, when the angular velocity is low and the developed torque is high (6–8 s), the electric power consumption is much greater than the mechanical output power. Finally, when the motor produces a stall torque (0–2, 4–6, –10 s), the mechanical power is zero, whereas the electric power is non-zero. Obviously, the demands for electric power are higher for higher torque requirements [see Eq. (17)]. The discussed results show that—for a frictionless reducer and different operation conditions—the efficiency of an electric drive system can vary from 0 to almost 1.

5 Conclusions

A simple model of friction effects in reduction gears is presented in the chapter. It may be used to account for friction in various types of reducers. The torques required to drive the system may be estimated prior to building the hardware prototype, and even prior to selection of gear type. The presented model of a reducer can work in two modes of simulation. In the "kinematic" mode vehicle motion is a result of imposing driving constraints. In the "dynamic" mode electro-mechanical properties of motors are taken into account, and motion results from supplying the input voltage. Essential differences between these two modes were discussed.

The virtual model of the Small Mobile Robot was tested; the study was focused on the movable track holders actuation system. The performed tests show that accounting for gear friction is crucial for proper estimation of torque requirements. It is also shown that in the case of using self-locking (irreversible) gears the calculated driving torques are completely different from the torques calculated for frictionless power transmission. Requirements for mechanical and electric power were calculated alongside with demands for driving torque.

The results of simulations show that—in the case of the movable track holders—it is desirable to apply speed reducers with friction level slightly above the self-locking limit. During the SMR operation the track holders are not going to rotate very often, thus it is crucial to keep them blocked without any power consumption. That is why speed reducers should be irreversible. On the other hand, when the track holders rotate, the power requirements are higher for higher friction levels, thus there is no point in pushing friction far beyond the self-locking limit.

The presented model supports the robot designer with quantitative results. Other typical robot manoeuvres, fairly different from the presented ones, can be studied using the same multibody model. An experimental verification of the simulated results is planned in the future.

Acknowledgments The project was co-financed by the European Regional Development Fund within the framework of the 1. Priority axis of the Innovative Economy Operational Programme, 2007–2013, through grant PO IG 01.02.01-00-014/08-00, and by the Institute of Aeronautics and Applied Mechanics statutory funds.

References

Frączek J, Wojtyra M (2011) On the unique solvability of a direct dynamics problem for mechanisms with redundant constraints and Coulomb friction in joints. Mech Mach Theory 46:312–334

Holmes DG, Lipo TA (2003) Pulse width modulation for power converters: principles and practice. Wiley-IEEE Press, New Jersey

Hughes A (2006) Electric motors and drives: fundamentals, types and applications, 3rd edn. Newnes, Oxford

Krause PC (1986) Analysis of electric machinery. McGraw-Hill, New York

Leonesio M, Bianchi G (2009) Self-locking analysis in closed kinematic chains. Mech Mach Theory 44:2038–2052

Ogata K (2009) Modern control engineering, 5th edn. Prentice Hall, New Jersey

Olędzki AA (1995) Modeling and simulation of self-locking drives. Mech Mach Theory 6:929–942

Siciliano B, Sciavicco L, Villani L, Oriolo G (2008) Robotics: modelling, planning and control. Springer, London

Slättengren J (2000) Utilization of ADAMS to predict tracked vehicle performance. SAE 2000 World Congress, Detroit, MI, USA

Wong JY (2008) Theory of ground vehicles, 4th edn. Wiley, New Jersey

MSC.ADAMS + ATV_2008r1 (2008) Documentation and help

http://www.projektproteus.pl/?lg=_en (2012) Accessed 4 June 2012

New Low-Cost Video Camera Pointing Mechanism for Stratospheric Balloons: Design and Operational Tests

Jarosław Jaworski, Kamil Bobrowski, Łukasz Boruc, Krzysztof Gedroyć, Krystyna Macioszek and Tomasz Rybus

Abstract Quality of video recordings from stratospheric balloons is poor due to motions of the balloons gondolas. This chapter presents new low-cost camera stabilization and control system (pointing mechanism) that is designated specifically to be used on the stratospheric balloons. The system design is described and the results of test-flight of the prototype of the proposed system are shown.

1 Introduction

Stratospheric balloons equipped with video cameras can be used for various terrain observations, environment monitoring or in crisis management. Cameras could be mounted as a secondary payload on balloons sent for investigations of the atmospheric conditions or with other scientific equipment and, therefore, increase mission's output. Various studies show that remote sensing performed from platforms operating in the stratosphere has some advantages in comparison to satellite observations (e.g., Everaerts et al. 2004) and such altitudes are only

J. Jaworski (✉) · K. Bobrowski · Ł. Boruc · K. Gedroyć · K. Macioszek
Students Space Association, Warsaw University of Technology, Warsaw, Poland
e-mail: scope2.0.team@gmail.com; imperator.jarek@gmail.com

K. Bobrowski
e-mail: scope2.0.team@gmail.com

Ł. Boruc
e-mail: scope2.0.team@gmail.com

K. Gedroyć
e-mail: scope2.0.team@gmail.com

K. Macioszek
e-mail: scope2.0.team@gmail.com

T. Rybus
Space Research Centre of the Polish Academy of Sciences, Warsaw, Poland

J. Sąsiadek (ed.), *Aerospace Robotics*, GeoPlanet: Earth and Planetary Sciences,
DOI: 10.1007/978-3-642-34020-8_8, © Springer-Verlag Berlin Heidelberg 2013

reachable by High Altitude Long Endurance Unmanned Aerial Vehicles (HALE UAVs) or by aforementioned stratospheric balloons.

In case of the stratospheric balloons, there are problems with smooth video recordings from video cameras due to motions of the gondolas. During the flight, the balloon's gondola moves and rotates in uncontrolled manner which makes observations of the Earth's surface very difficult. Another important matter is the ability to turn the camera in the desired direction, which would allow it to make continuous observations of chosen areas on the ground not directly below the vehicle, as balloon route cannot be controlled.

Commercially available camera stabilization systems for UAV applications are very expensive and are not entirely suited for stratospheric balloons (i.e., the required ability to compensate unlimited rotations of the gondola around the vertical axis, the ability to work on high altitudes). While engineering community and numerous companies are focused on complicated, multi-applicable and highly accurate systems which are also capable of vibration dumping [e.g., SPEKTROP project (Grygorczuk et al. 2010)], we propose a simple system made from COTS components which is designated specifically to be used on the stratospheric balloons. This three DOFs video camera stabilization and control system (pointing mechanism) is capable of compensating changes in balloon's gondola attitude and allows camera to track selected ground targets during the flight. Prototype of such system was constructed and tested under the Stabilized Camera Observation Platform Experiment (SCOPE) as a part of the Balloon Experiments for University Students (BEXUS) programme.

The chapter is divided into six sections (including the introduction). In the Sect. 2, environmental constraints and system objectives are described. In the Sect. 3, design of the camera stabilization and control system is presented. Section 4 contains description of pre-flight tests and in-flight verification of the prototypic realization of this system performed during BEXUS 11 flight. The conclusions are presented in Sect. 5. Section 6 contains description of future work.

2 Environmental Constraints and System Objectives

Video recordings from stratospheric balloons flights BEXUS 6/7 and BOBAS 3, as well as quantitative data of gondola's attitude obtained from the Low.Co.I.N.S. experiment [Low Cost Inertial Navigation System tested during BEXUS 6 flight (Buccilli et al. 2009)] were used to identify the nature of gondola's motion during the flight and to define requirements for motor velocities of the stabilization system. While there are no significant vibrations during balloon's flights, changes in gondola's attitude might be fast. The maximal gondola's angular velocities recorded were: during ascent -10 deg/s, level flight -1 deg/s and descent -60 deg/s for medium-weight (about 100 kg) gondola during BEXUS 6 flight in steady weather conditions. For heavier gondolas and larger balloons, those velocities are expected to be significantly smaller (Fig. 1).

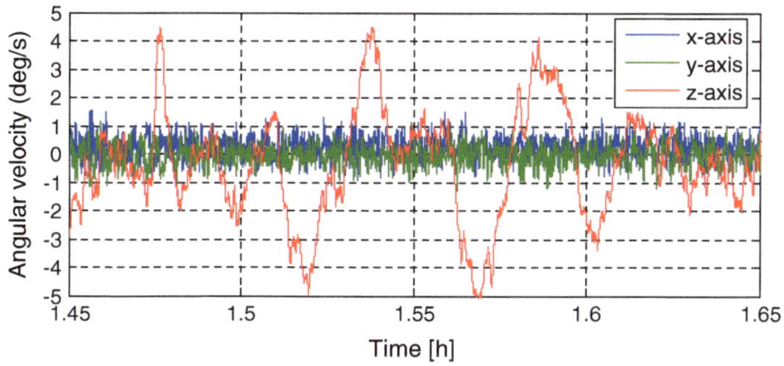

Fig. 1 Angular velocities of the stratospheric balloon gondola during ascent (based on Low.Co.I.N.S. data obtained during BEXUS 6 flight)

It is also important to note that rotations around the vertical axis are infinite and slip ring must be used on the last joint. The presented camera stabilization system was designed to compensate changes in gondola's attitude during the first two phases of a similar flight and to withstand 10 g vertical acceleration in the final phase at the end of flight. Another driving factor for the design were the extreme environmental conditions at the maximal height reached by the balloons (35 km): temperature below −70 °C and pressure below 1 mbar, resulting in almost no convection.

Apart from the educational aspects of this student project, the main objective of SCOPE was a prototypic realization of low-cost video camera stabilization and control system and in-flight verification and assessment of its performance. This prototype is a slightly simplified version of the proposed design: no slip ring was used on the third joint and, therefore, rotations of this joint are limited.

3 System Design

3.1 General Overview

The main element of the system, mounted to the structure, is called the Platform Orientation Mechanism (POM). POM consists of three frames connected by rotational joints actuated by stepper motors. A video camera is mounted to the last joint. Motion of joints is provided by three motor drivers, controlled by the Interface Board (IB). Encoders' data is used as a feedback signal. PC/104 is used as the On-board Computer (OBC) and it is responsible for gathering data, computations and communication with the Ground Station. An electronic box and power supply system with a battery box are mounted on the structure above POM. The electronic box (containing: electronic boards, PC/104 and power control subsystem) and the battery box are mounted on the structure above POM. The GPS

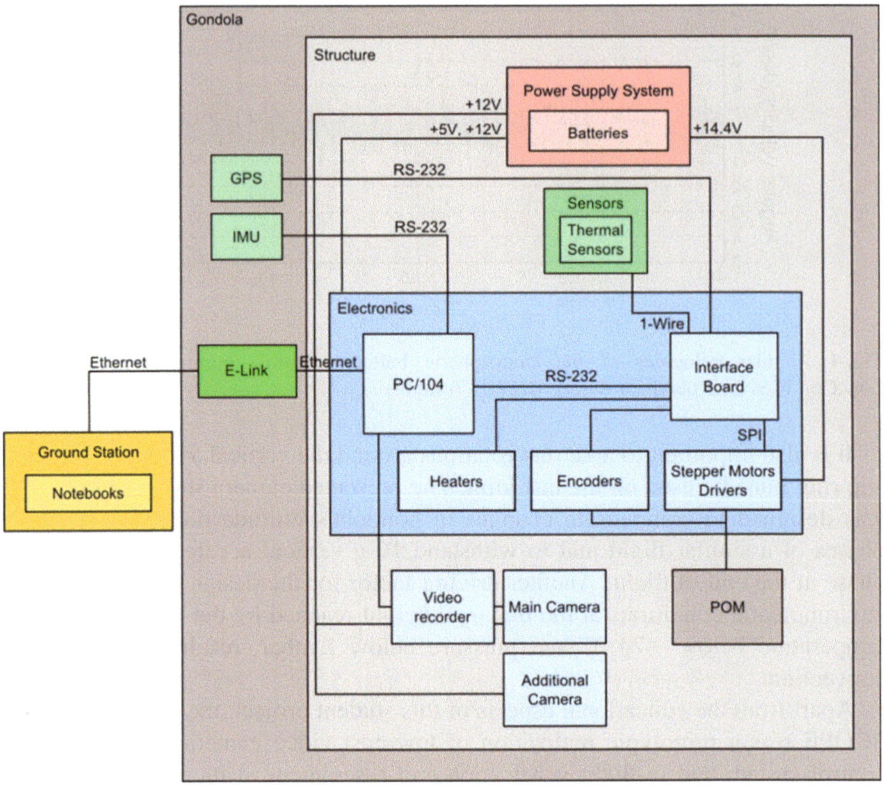

Fig. 2 Overview of the camera stabilization and control system

receiver gives information about gondola's position. The Inertial Measurement Unit (IMU), which is mounted on a long boom outside the gondola to minimize electrical disturbances, provides actual gondola's attitude. Basing on this data, PC/104 computes joints' positions that are needed to achieve the desired camera orientation. The data flow between PC/104 and other sensors (except IMU) is carried out via IB. No antenna is used and communication must be provided externally (in case of BEXUS flight it was E-link system belonging to balloon's providers—Esrange). The schematic overview of the system is presented in the Fig. 2 and the basic parameters of the prototype tested during BEXUS 11 flight are shown in the Table 1.

3.2 Control Software and Algorithms

Software running on the PC/104 is responsible for: controlling the entire experiment, receiving telecommands from the Ground Station (GS), sending telemetry to

Table 1 Basic parameters of the camera stabilization and control system

Total mass	21.87 kg
External dimensions	466 × 550 × 538 mm
Camera maximal deflection from vertical axis	45°
Rotations around vertical axis	Unlimited (limited to 310° in the prototype)
Compensated changes in gondola's attitude	30 deg/s
Camera resolution	540 × 576 px
Operational time (available power)	12 h
Minimal operational temperature (confirmed)	−70 °C
Minimal operational pressure (confirmed)	1 mbar

GS, gathering data from sensors (IMU, GPS and thermal sensors) and storing this data on on-board memory, computing camera attitude that should be obtained in order to point toward selected target, and sending computed joints positions to IB. The source code of PC/104 control software is written in C++ and runs on Linux OS.

On the basis of gondola's current position (from the GPS data) and selected target's geographical coordinates, the Target Tracking Algorithm (TTA) calculates the desired camera attitude for pointing to the target. Then, taking into account gondola's movements, the Joints Positions Determination Algorithm (JPDA) on the basis of the desired (computed by the TTA) and gondola's actual attitude (measured by the IMU) calculates joints' positions that would keep the camera pointing to the target. Working of these two algorithms is shown in the Fig. 3. Various modes of operation can be used during the flight. Beside pure stabilization and tracking of ground targets, the system can be used to perform pre-programmed sequence of motion (e.g., taking panorama or capturing terrain mosaic) or an operator can manually change the desired orientation of the camera by telecommands.

3.3 *On-Board Computer and Electronics*

The main element of on-board electronics is Kontron MOPSlcdLX AMD LX800 PC/104 connected to the IB, on which ATmega128 microcontroller is located. This microcontroller gathers signals from sensors (positions of POM's joints,

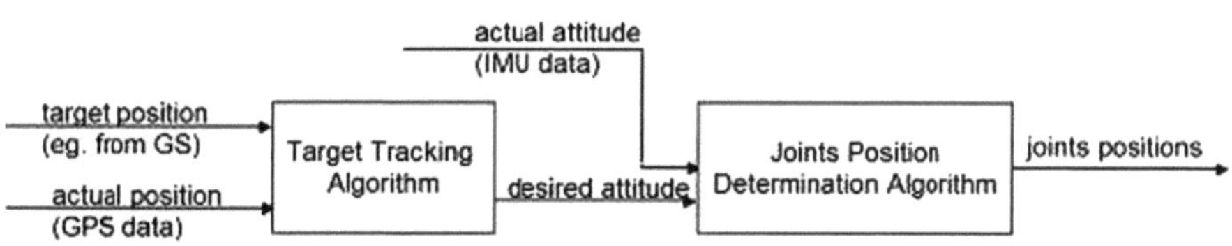

Fig. 3 Control algorithms for tracking ground targets

temperatures, geographical coordinates). RS-232 communication interface is used to exchange data between the IB and PC/104. Incremental encoders (connected to motor drivers) and motors of two first stages are working in the feedback loop (with PID based regulator) to make POMs move faster and more precisely. To achieve high quality and smooth shaft rotation of the stepper motors (1/256 microstep), specialized motor controllers were designed (with Atmega8 microcontroller).

3.4 Sensors

LC-1/3 Sony 520TVL board analog camera with 520 TVL resolution and Sony CCD 1/3" sensor was used in the prototypic realization of the camera stabilization and control system (this camera can be seen in the Fig. 4). The main advantage of this camera is low weight (7 g) and small size (30 × 30 × 20 mm excluding optics). A LCMTV 25 mm lens was used. The camera was mounted on a board frame grabber A/V which allows recording with resolution of up to 1,280 × 720 pixels and up to 30 frames per second. Recordings were stored on a micro SDHC card. Dimensions of this system are very small: the main board of 60 × 60 mm and SDHC board of 50 × 36 mm. The main advantage of this system is that it is autonomous and that it can work even if any malfunctions of other subsystems occur. On the other hand, recordings were not available on GS during the flight. Moreover, this solution allows using a slip ring.

To stabilize the camera in the desired orientation, data of actual gondola's orientation is provided by IMU (MicroStrain 3DM-GX1). It has nine sensors: three accelerometers, three gyros and three magnetometers. All of them are provided with full temperature compensation which highly increases accuracy of its readings. High accuracy assured by the producer (± 0.5° for static and ± 2° for dynamic conditions) is extremely important in the proposed stabilization system. Garmin GPS 18x LVC GPS receiver was chosen, because it can work above the altitude of 60 kft (approximately 18.2 km), while majority of commercially available GPS receivers cannot work above this level.

Fig. 4 Video camera mounted on the POM

3.5 Mechanical Design

The main element of the stabilization system, i.e., the Platform Orientation Mechanism (POM), is mounted to the aluminium structure. The upper part of the structure, separated from POM by an internal panel, holds electronics box and battery box. Glass plate is mounted below POM for safety reasons. The overview of the structure is presented in the Fig. 5.

POM is divided into three aluminium-made stages. Moments of inertia for particular stages are very small and counterweight is used on the last stage for balance. POM's motion is realized by three stepper motors. Encoders provide rotation data about shafts position to improve control. Transmissive Optical Sensors are used to determine the resting POM position (camera pointed vertically downwards), because selected encoders do not provide an absolute position. Schematic view of POM is presented in Fig. 6, while pictures of the prototypic realization of the designed system are presented in Fig. 7.

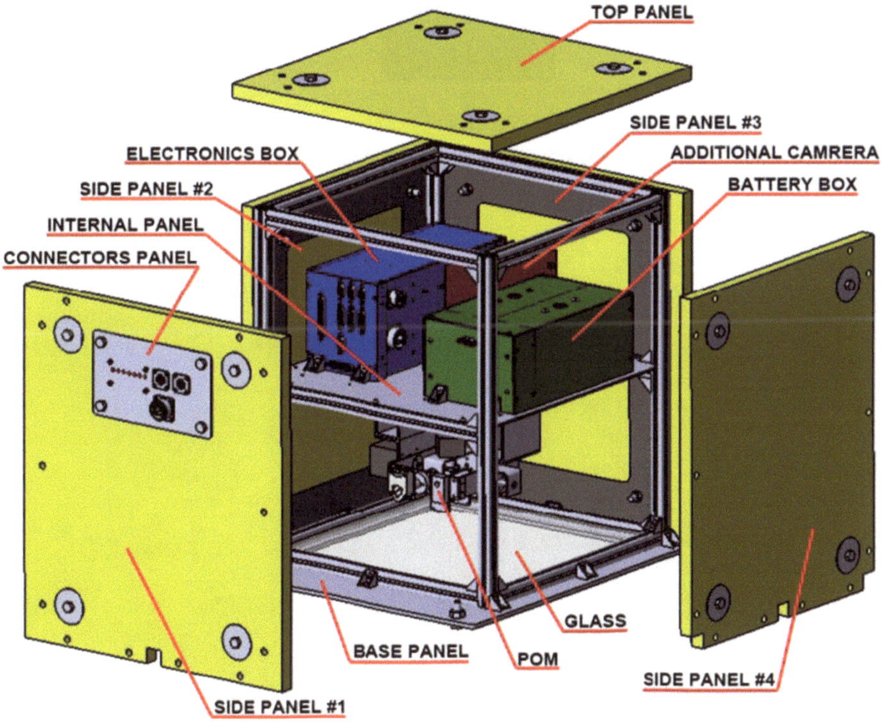

Fig. 5 Schematic view of the prototypic realization of the camera stabilization system

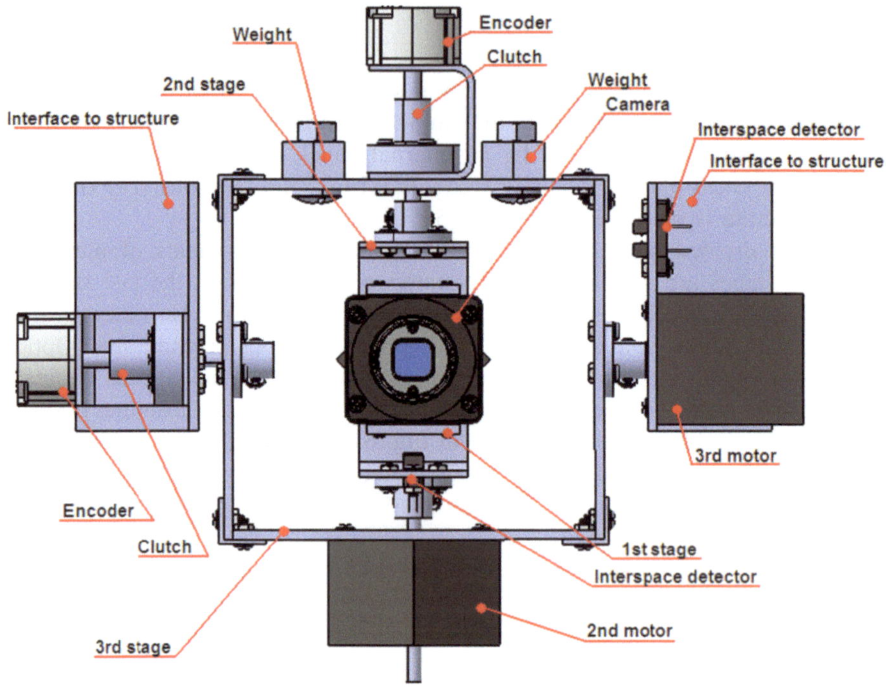

Fig. 6 Schematic view of the Platform Orientation Mechanism (POM)

Fig. 7 Picture of the entire system (*left*) and POM (*right*)

4 Operational Tests and Verification of the Proposed Solution

The prototype of the proposed system was constructed with one important simplification: no slip ring was used on the third joint due to problems with camera connection. An additional small video camera was used to record POM's movements. This allows continuous observations of system's work and may allow identifying the source of malfunction in case of a failure.

Preliminary tests of the control algorithms (TTA and JPDA) were done by computer simulations of the stabilization system. The detailed thermal analysis was performed (see Fig. 8) to design thermal control system which will ensure that during the entire flight temperatures will be within the range defined for all components (−40–70 °C for the electronic box, 0–40 °C for the battery box). It turned out that passive thermal control system consisting of Styrofoam insulation panels is sufficient and such system was used. Tests in thermal-vacuum chamber were performed (see Fig. 9) in order to confirm that system's electronics can operate in specified temperature and pressure range. These tests proved that system works well even in the wider range of parameters: temperature −70 °C and pressure 1 mbar.

The prototype of the proposed camera stabilization and control system was successfully tested in September 2010 during reduced on-ground tests. Those tests proved that the selected concept of the system is correct and 2-axis stabilization works satisfactorily, but some problems were detected in compensation of motions around vertical axis due to lower than expected accuracy of IMU. Final in-flight

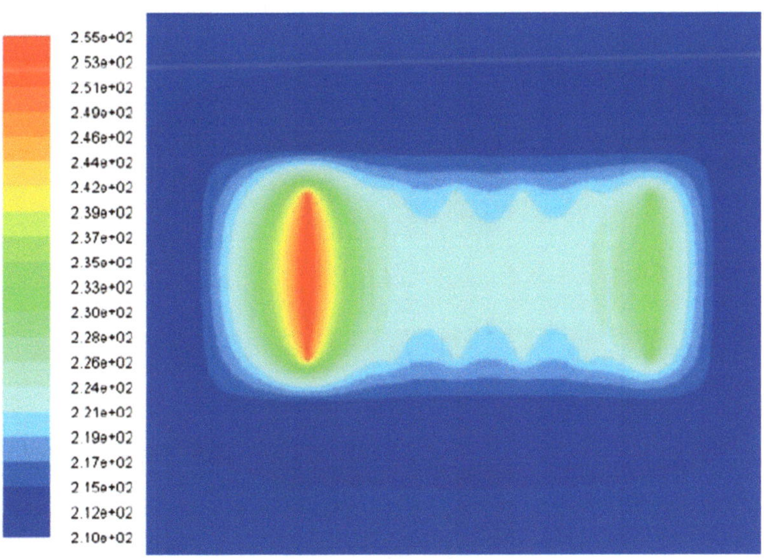

Fig. 8 Results of thermal analysis

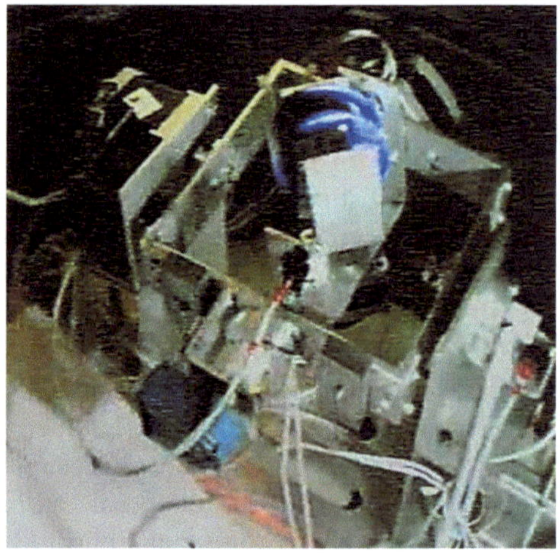

Fig. 9 Tests in thermal-vacuum chamber

test took place during BEXUS 11 flight from Kiruna (Sweden) on the 23rd of November 2010 (Fig. 10). Although not entirely successful due to problems with on-board software and communication with the Ground Station, this flight proved that the constructed system can work in every phase of the flight, withstand landing and operate without disturbing other systems on-board the gondola. However, flight preparations showed that the designed system should be more ergonomic. Quality of the video recorded during the entire flight by the main camera is very good and allows recognition of many surface features (see Fig. 11). Video recording from the additional camera shows that POM was nominally performing pre-programmed test sequence of motion at all altitudes (see Fig. 12). During such sequence all joints are moved to extreme positions with the maximal velocity.

Fig. 10 BEXUS 11 stratospheric balloon just before launch

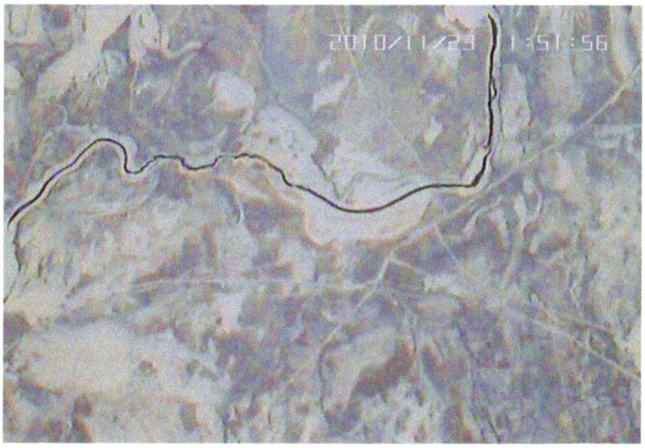

Fig. 11 View from the main camera taken at the altitude of 27,200 m

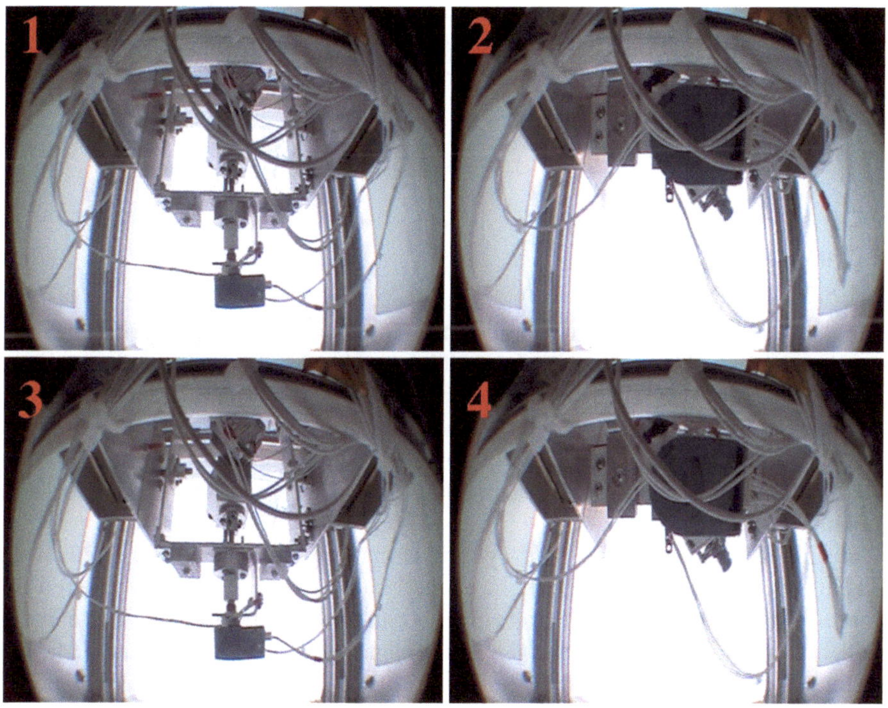

Fig. 12 Pre-programmed test sequence of motion performed at the height of 16,000 m (pressure: 93 mbar, external temperature: −53 °C)

5 Conclusions

The new camera stabilization and control system designed specifically for stratospheric balloons was constructed and thoroughly tested. Innovative design allows high camera deflections from vertical axis (up to 45°) and unlimited rotations around vertical axis. The ability to track ground targets selected by an operator during the flight may significantly enhance possibilities to use stratospheric balloons for various ground observations. The design based on COTS components ensured low price of the prototype—its overall cost was below 5,500 EUR. This cost could be even lower, because during construction of the prototype many components recommended by the European Space Agency (specified for harder conditions than those at the height of 35 km) were used.

The final test flight on the BEXUS 11 (Balloon Experiments for University Students) stratospheric balloon was not entirely successful, but proved that selected concept of the pointing mechanism is correct and that system can work in extreme conditions at the height above 30 km. This flight also allowed recognition of certain issues (e.g., software problems) that must be resolved. It should also be noted that in cases when balloon's gondola is very heavy and flight is smooth (as it was in the case of BEXUS 11), changes in gondola's attitude are slow and only compensation of rotational motion around vertical axis becomes important. However, during such flights, the ability to point camera in the desired direction is still very important.

6 Future Work

The designed system, after some minor modifications and further miniaturization, can be used on Unmanned Aerial Vehicles (UAVs) for various observations. Infrared camera can be used instead of visible light camera. One could also consider using the proposed system as a pointing mechanism for a directional antenna, which needs to be precisely pointed in the direction of a ground station. Such solution might be useful for balloon-borne experiments that transmit large amounts of data.

Quantitative measurements of system's accuracy are to be conducted in the coming months. The next test flight of the constructed system is considered onboard of one of experimental stratospheric balloons flown in Poland by teams of students and amateurs or on-board a small plane. It is also possible that dedicated balloon project SKAXUS will be performed by the Students' Space Association at the Warsaw University of Technology and the proposed camera stabilization and control system could be used as a payload.

Acknowledgments The described project was conducted as a part of the REXUS/BEXUS programme for the university students. This programme is realised by the German Aerospace Center (DLR) and the Swedish National Space Board (SNSB) in which the Swedish share of the payload has been made available to students from other European countries through collaboration with the European Space Agency (ESA). Project was partially financed by the ESA PECS project no. 98089 and by Warsaw University of Technology scientific grants for students. The authors would like to thank engineers and scientists working in the Faculty of Power and Aeronautical Engineering of the Warsaw University of Technology and Space Research Centre of the Polish Academy of Sciences for their help and support. The authors would also like to acknowledge kosmonauta.net for active promotion of SCOPE project in Poland.

References

Buccilli T, Folina A, Medaglia E, Montefusco P, Oliva M, Palmerini G, Sestito A (2009). A low cost inertial navigation experiment onboard balloons. In: Proc. of the 19th ESA symposium on european rocket and balloon programmes and related research, Bad Reichenhall, Germany

Everaerts J, Claessens M, Biesemans J, Lewyckyj N (2004). Pegasus: high resolution remote sensing from a stratospheric unmanned aerial vehicle. In: Proc. of the 20th international society for photogrammetry and remote sensing congress,vol 35. Istanbul, Turkey (IAPRS)

Grygorczuk J, Juchnikowski G, Wawrzaszek R, Seweryn K, Dobrowolski M, Sobolewski M, Przybyła R, Rataj M, Orleański P, Gut H, Ciesielska M, Koruba Z (2010) Stabilization of the multispectral imaging system for UAV: GT3 task for the project 'SPEKTROP'. Scientific Aspects of Unmanned Mobile Vehicles, Kielce

The Dynamics Influence of the Attached Manipulator on Unmanned Aerial Vehicle

Grzegorz Chmaj, Tomasz Buratowski, Tadeusz Uhl, Karol Seweryn and Marek Banaszkiewicz

Abstract The chapter describes the development and operation of Unmanned Aerial Vehicle (UAV) type flying robot with attached manipulator. The hardware, software architecture and mathematical description of the system used to control the robot is presented. The results of test rigs connected with flying the robot with attached manipulator have been presented and discussed.

1 Introduction

Rapid improvement of technological development in last three decades has led to progress in development of UAV projects. Originally, researchers had been developing UAVs based on fixed-wing models. This type of design is simpler to control as opposed to UAV based on a helicopter model. The fixed-wing model approach to UAV design was adopted by many research centers worldwide (Bar-Itzhack and Harman 2002; Doitsidis et al. 2004; Tsai 1999; Sasiadek et al. 2002), due to the fact that control systems were more developed and easier to apply in this type of advanced mechatronic solutions. For example, a single GPS receiver may act as both the position and velocity sensor (Wood and Kennedy 2003). Along with technological development, research centers started to move their experience onto a more complex model—a helicopter. Development of more UAV projects based on helicopter models called for rules to classify this kind of projects. One of standards used to classify UAVs was introduced by UVS-International. Table 1 shows the

G. Chmaj · T. Buratowski (✉) · T. Uhl
AGH University of Science and Technology, Cracow, Poland
e-mail: tburatow@agh.edu.pl

K. Seweryn · M. Banaszkiewicz
Space Research Center of the Polish Academy of Science, Warsaw, Poland

J. Sąsiadek (ed.), *Aerospace Robotics*, GeoPlanet: Earth and Planetary Sciences,
DOI: 10.1007/978-3-642-34020-8_9,

Table 1 Extract of UAV categories defined by UVS international

No	Category name	Mass (kg)	Range (km)	Flight altitude (m)	Endurance (h)
1	Micro	<5	<10	250	1
2	Mini	<25/30/150	<10	150/250/300	<2
3	Close range	25–150	10–30	3,000	2–4
4	Medium range	50–250	30–70	3,000	3–6
5	High altitude long endurance	>250	>70	>3,000	6

extract from that standard. The flying robot described in this chapter belongs to the 'Mini' category.

The UAV application was developed in the Department of Robotics and Mechatronics UST. It was design for diagnostic purposes. The main aim of this project is to allow a ground operator to carry out diagnostic tasks, taking advantage of both the mobility of flying robot and an onboard vision system or other diagnostic devices. Moreover, ground operator has to have control over current flight path and diagnostic tasks: experiment planning and diagnostic task execution. The assumption is that the experiment planning will be connected with flight trajectory planning (before take-off) and the diagnostic task execution will be used to test an object of diagnostic (using the mobile platform).

1.1 UAV Robot Description

The main purpose of the flying robot is to carry out diagnostic tasks using video system or other devices, e.g., manipulator. The use of the additional devices attached to the robot, during the flight, requires very stable operation and stationary hovering. Such applications require a typical mechatronic design approach where all three aspects: mechanics, electronics and informatics, should be in synergy. The mechanical part of the robot is based on SST-Eagle2-GS Long Tail helicopter model produced by HIROBO, presented in Fig. 1.

For this model, the main rotor diameter equals 1.8 m. The helicopter is powered by a single cylinder, 2-stroke, 2.7 HP combustion engine. The take-off weight with a fully equipped system is approximately 10 kg. The frame of the helicopter, including avionics, battery and fuel, weigh around 7 kg. Units equipped with electric drive are characterized by low levels of mechanical vibrations, mainly generated by the drive unit itself. This feature of electric units is a key parameter in this type of solutions. It directly affects the quality of data gathered from the Internal Measurement Unit (IMU), which is the primary source of information about the position and orientation of the robot in space. In a situation where the robot is equipped with a manipulator, low levels of mechanical vibrations are directly affecting the quality of positioning of the manipulator arm. The disadvantages of the electrical engine are small capacity and short operation time. Taking into account one of the project

Fig. 1 Unmanned aerial vehicle

objectives, which stated that we have to study the behavior of the robot equipped with a manipulator, it was decided that in order to increase the capacity, the robot must be equipped with a combustion engine. The selected model is powered by an internal combustion engine and has a load-carrying capacity equal to the weight of the unit. It should be noted that this type of commercially available models of helicopters may be categorized in four classes: 30, 60, 90 and 120. The selected type of helicopter is in the largest class—120.

The electronics of the robot is directly connected with algorithms necessary to control the system. The modern flight control system is a combination of several sensors, Global Navigation Satellite System, software and a control system designed to control the aircraft actuators. The primary task of this system is the estimation of the position (latitude, longitude and altitude), velocity (in a northerly direction, in an easterly direction, floating and sinking) and orientation (flight direction, lean, tilt). In order to estimate those parameters it is necessary to integrate into one large system subsystems as follows: Inertial Measurement Unit system (IMU is a combination of gyro sensors and accelerometers used to measure angular velocity and acceleration around proper axes X, Y, Z and along Cartesian coordinates), Attitude and Heading Reference System(AHRS is a combination of the IMU with Earth's magnetic field sensor providing information about the direction of flight), GNSS system (provides information about the position, speed and altitude). The use of GNSS in the Inertia Navigation System (INS) can be implemented in two configurations: loosely coupled INS/GPS, tightly coupled INS/GPS. The difference between those configurations is that in the system configuration with loosely coupled INS with GPS, in case of less than four satellites visibility, the most widely used Kalman filter (KF) (Zhang et al. 2005) is unable to adequately calculate a good estimation of the position and speed of the aircraft, while in the tightly couple configuration even associated with a single satellite data are taken into account in the process of the Kalman filter calculation (Haug 1989; Heffley and Mnich 1988; Mao et al. 2004; Merwe and Wan 2004).

The flight control systems can be categorized in two groups where flight control system is operating in an open loop or in a closed loop. The first group includes all civilian uses, where the pilot always makes the final decision. The second group includes military applications and the applications connected with UAVs, where the final decision is always made by the flight control system.

In presented solution, the robot was equipped with INS system built from commercially available components (Fig. 2) integrated in autonomous electronic system MP2128Heli produced by Micropilot. The system weighs just 28 g and its dimensions are 100 × 40 × 20 mm. This system allows automatic vertical take-off and landing, has the possibility of self-programming during the flight, communication with terrestrial base station, full configurability of the all system parameters. Besides the INS electronic components, the system is equipped with ultrasonic sensor, sensor measuring the fuel level in the tank, sensors measuring engine temperature and engine speed and wireless communication. The core of MP2128Heli unit is a Reduced Instruction Set Computers (RISC) microprocessor. The above-mentioned electronic unit has a built-in GPS receiver, model TIM 4P produced by u-blox company, working at a frequency of 4 Hz. Information about the orientation of the helicopter are fed into the MP2128Heli system based on the build-in IMU system, while the primary source of information on the position of the robot is a GPS receiver.

The INS system uses two 6-state Kalman filters algorithms. These filters work with 200 Hz sampling rate. One of them is integrated with a GPS receiver in the loosely coupled configuration.

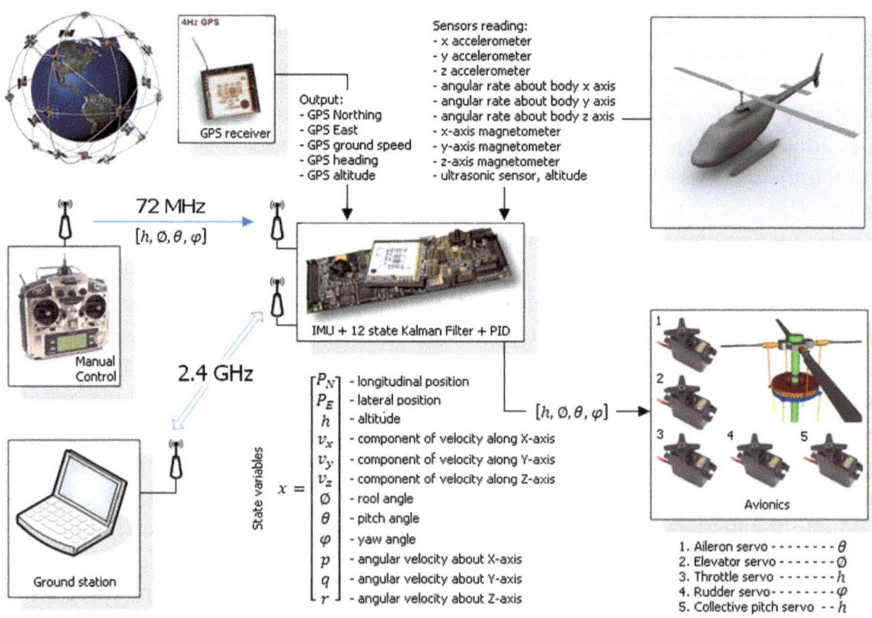

Fig. 2 Infrastructure of on-board flight control system and its components

The general task of Kalman filter in Micropilot system is to estimate the errors associated with the position, velocity and orientation. Errors appear due to the inaccuracy of sensors. In commercial INS/GPS systems, state vector size is as high as 50. However, we should take into account that the computing power needed to operate a Kalman filter (Jang and Tomlin 2003) is the third power of the number of state variables, and the size of estimated states is as accurate as the mathematical model of the dynamics of the system (Jensen et al. 2004; Hald et al. 2005). Therefore, in INS/GPS system for UAV applications, where computational power is limited, the state vector size is reduced to between 12 and 17. In the MP2128Heli system, the two implemented 6-state Kalman filters (Hald et al. 2005) are responsible for the estimation of orientation and position states of the robot. Vectors of these filters are listed below:

- state vector that stores information about the orientation of the robot:

$$q_1 = [\theta, \varphi, \psi, \theta_{bias}, \varphi_{bias}, \psi_{bias}] \quad (1)$$

where:

θ	rotation around the X axis
φ	rotation around the Y axis
ψ	rotation around the Z axis
θ_{bias}	systematic error for angular velocity of the gyroscope about the X axis
φ_{bias}	systematic error for angular velocity of the gyroscope about the Y axis
ψ_{bias}	systematic error for angular velocity of the gyroscope about the Z axis

- state vector that stores information about the position of the robot:

$$q_2 = \left[P_x, P_y, P_z, V_x, V_y, V_z\right] \quad (2)$$

where:

P_x	position in an northerly direction (direction X)
P_y	position in an easterly direction (direction Y)
P_z	position in vertical direction (direction Z)
V_x	velocity in a northerly direction (direction X)
V_y	velocity in an easterly direction (direction Y)
V_z	velocity in vertical direction (direction Z)

The directions X, Y, Z in the description of Eqs. (1) and (2) form the axes of Cartesian coordinate system associated with the frame of the robot shown in Fig. 3. One of the project tasks will be to fine-tune the described flight control system of autonomous helicopter equipped with a manipulator, in which the

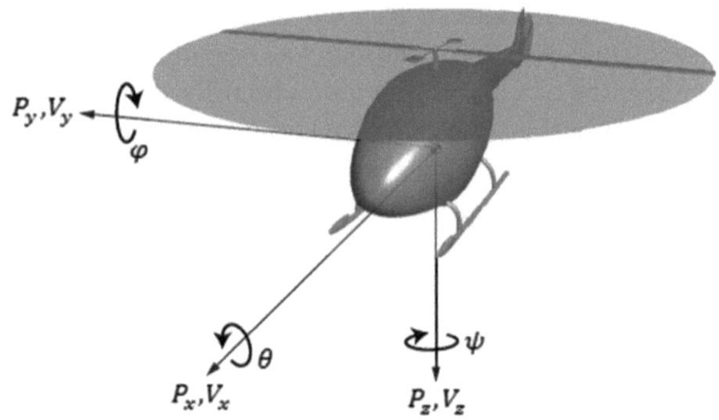

Fig. 3 The definition of the coordinate system associated with the robot's frame

motion of the manipulator will change the center of balance of the system, which significantly affects the parameters of flight control.

2 Mathematical Description of the Model

Model of dynamics of the helicopter is based on the simplified model presented by Hald et al. (Kalman 1960). The following equations are used for computation of forces and torques acting on the helicopter's center of gravity (CG):

$$\mathbf{F} = \begin{bmatrix} f_x \\ f_y \\ f_z \end{bmatrix} = \begin{bmatrix} -T_{MR}\sin(\beta_{1c}) - \sin(\theta)mg \\ T_{MR}\sin(\beta_{1s}) + T_{TR} + \sin(\phi)\cos(\theta)mg \\ -T_{MR}\cos(\beta_{1s})\cos(\beta_{1c}) + \cos(\phi)\cos(\theta)mg \end{bmatrix} \tag{3}$$

$$\tau = \begin{bmatrix} L \\ M \\ N \end{bmatrix} = \begin{bmatrix} f_{y,MR}h_m - f_z y_m + f_{y,TR}h_t + Q_{MR}\sin(\beta_{1c}) \\ -f_x h_m - f_z l_m - Q_{MR}\sin(\beta_{1s}) \\ f_x y_m + f_{y,MR}l_m - f_{y,TR}l_t + Q_{MR}\cos(\beta_{1c})\cos(\beta_{1s}) \end{bmatrix} \tag{4}$$

where T_{MR} is the main-rotor thrust, T_{TR} is the tail-rotor thrust, β_{1s} is the lateral flapping angle, β_{1c} is the longitudinal flapping angle, θ is the Euler angle of helicopter pitch, ϕ is the Euler angle of helicopter roll, m is the helicopter mass, g is the gravitational acceleration, l_m, y_m and h_m are the distances between the rotor hub and helicopter CG along the x, y and z axes, respectively, l_t is the distance from the center of the tail rotor to CG along z axis, Q_{MR} is the main rotor drag, $f_{y,MR}$ is the force along y caused by the main rotor and $f_{y,TR}$ is the force along y caused by the tail rotor. The main rotor thrust is described by the following equation (Welch and Bishop 2001):

$$T_{MR} = (\omega_b - v_i)\frac{\rho\Omega R^2 aBc}{4} \tag{5}$$

and the tail-rotor thrust is described by:

$$T_{TR} = \frac{f_{y,MR} l_m + f_x y_m - Q_{MR} \cos(\beta_{1c}) \cos(\beta_{1s})}{l_t} + u_{ped} \quad (6)$$

where ω_b is the wind velocity relative to the blade, v_i is the induced velocity, ρ is the density of the air, Ω is the rotational velocity of the main-rotor, R is the radius of the blade, a is the two-dimensional constant lift curve slope, B is the number of blades, c is the mean blade cord length and u_{ped} is the rudder control input.

The main-rotor drag is described by:

$$Q_{MR} = -\left(A_{Q,MR} T_{MR}^{1.5} + B_{Q,MR}\right) \quad (7)$$

where $A_{Q,MR}$ is the relationship between the main-rotor thrust and the drag, $B_{Q,MR}$ is the initial drag of the main rotor when the blade-pitch angle is zero. Constant values in Eqs. (3–7) are taken from the SST-Eagle 2-GS helicopter data, while other values are computed with the equations presented by Hald et al. (Kalman 1960).

Generalized equations of motion for the manipulator can be derived on the basis of second order Lagrange equations (Quigley et al. 2005) and take the following form:

Manipulator is modeled in SimMechanics as a chain of rigid bodies connected by joints (Fig. 4) (Kumar and Jann 2004). The geometrical and inertial properties of manipulator's links are obtained from CAD model of the manipulator. The model of the manipulator is connected to the rigid body of helicopter, and, as a result, both systems can dynamically interact (state of the helicopter can affect state of the manipulator and vice versa). In general, equations of this complex system can be expressed as:

$$\begin{bmatrix} \mathbf{M_H} & \mathbf{M_{HM}} \\ \mathbf{M_{MH}} & \mathbf{M_M} \end{bmatrix} \begin{bmatrix} \ddot{\mathbf{x}} \\ \ddot{\mathbf{q}} \end{bmatrix} + \begin{bmatrix} \mathbf{C_H} & C_{HM} \\ \mathbf{C_{MH}} & \mathbf{C_M} \end{bmatrix} \begin{bmatrix} \dot{\mathbf{x}} \\ \dot{\mathbf{q}} \end{bmatrix} + \mathbf{G} = \begin{bmatrix} \mathbf{F} \\ \tau \\ \mathbf{Q} \end{bmatrix} \quad (8)$$

where M_H is the mass matrix for the helicopter, M_M is the mass matrix for the manipulator, C_H is the Coriolis matrix for the helicopter, C_M is the Coriolis matrix for the manipulator, M_{HM}, M_{MH}, C_{HM} and C_{MH} express dynamic interactions between the helicopter and the manipulator, G is the vector of potential forces, x is the state vector of the helicopter as a rigid body (three Euler angles describing the orientation and three components of the position vector), q is the state vector of manipulator joints, F and τ are forces and torques acting on the helicopter [resulting from rotors dynamics and described by Eqs. (3) and (4)] and Q is the vector of control torques applied in the joints.

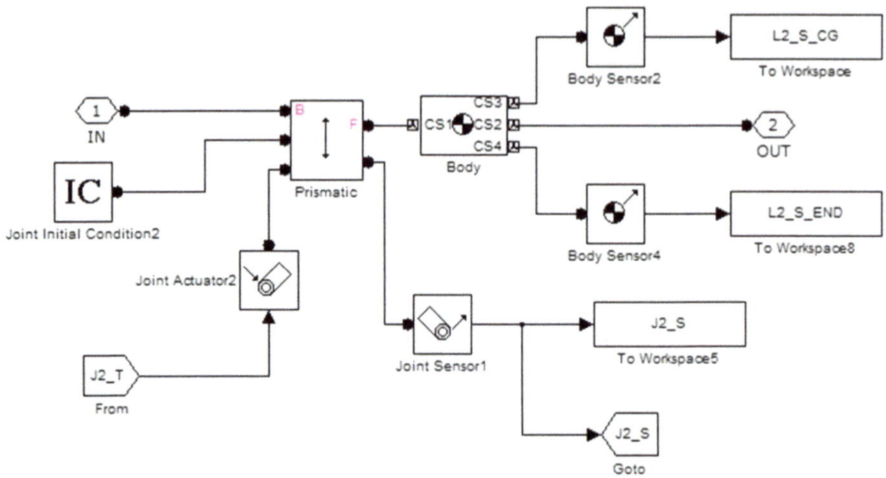

Fig. 4 SimMechanics model of the manipulator's second joint and link

3 Tests of the Autonomous Helicopter Equipped with Manipulator

Integrated prototype of autonomous helicopter with manipulator is shown in Fig. 5. For the first test flight, the manipulator arm has been extended to its limit and the manipulator was balanced and mounted in the center of gravity of the UAV. The aim of the first test flight was to examine the behavior of the UAV control system with equipped manipulator and to calculate the level of mechanical vibrations generated by air stream flowing over the manipulator arm, generated by a helicopter main rotor and generated by the helicopter engine. The objective of first test flight was to conduct an autonomous hovering at a height of 3 m. The first step of this test was to examine the influence of the added weight (manipulator) on

Fig. 5 UAV with attached manipulator

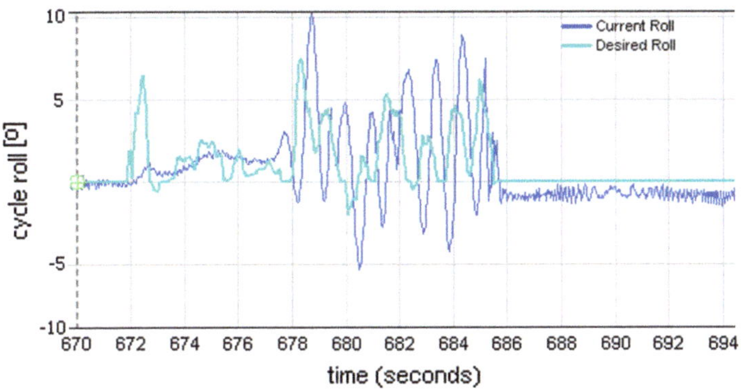

Fig. 6 Current and desired roll angle during unstable system operation

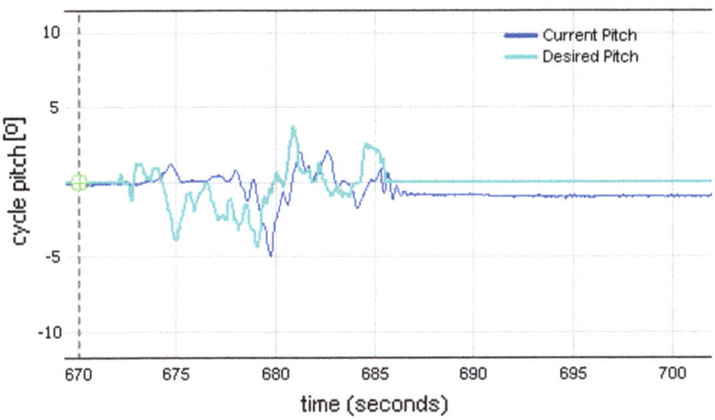

Fig. 7 Current and desired pitch angle during unstable system operation

the parameters of the helicopter's control system. Taking into account that the added weight has changed the total weight of the helicopter by 25 %, and that MP2128Heli is equipped with PID controllers, it was expected that a situation may occur in which the control system may exceed the limit of stability. The test confirmed the necessity to find algorithms that will adjust the setpoints of the PID controllers, due to the fact that shortly after takeoff the helicopter fell into lateral oscillations. Figures 6 and 7 show the waveforms of roll and pitch angles that are responsible for maintaining the helicopter in hovering, which were recorded during the test. Light color indicates the desired angles and dark blue color indicates the actual angles of roll and pitch of the helicopter. It can be seen from the drawings that the helicopter's roll angle started to oscillate, which indicates lack of stability of the control system. It is worth noting that the inclination angle (pitch angle) does not oscillate. This is caused by the mounting position of the manipulator namely; the mass of the manipulator has been displaced in the Y direction, resulting in an increase of inertia in the X direction.

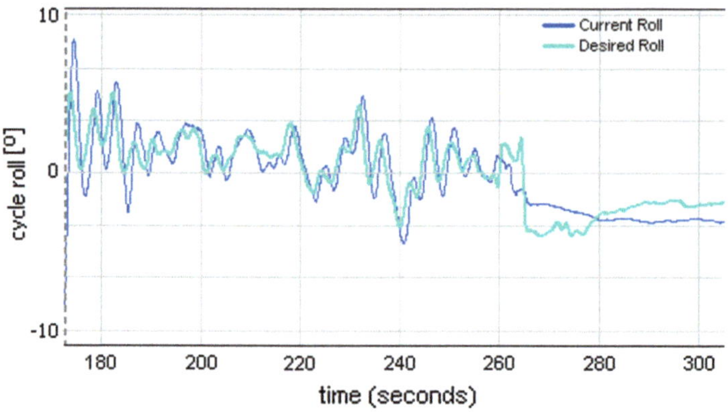

Fig. 8 Current and desired roll angle during stable system operation

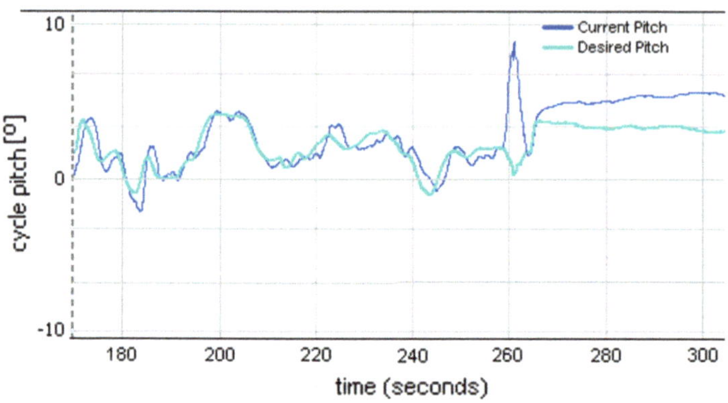

Fig. 9 Current and desired pitch angle during stable system operation

The following tests focused on tuning new PID setpoints. After several test flights the proper parameters of PID regulators, which ensure stable helicopter flight with additional mass, were determined. The time courses of the roll and pitch angles, for stable operation, are shown in Figs. 8 and 9.

After determining the control system parameters, further tests have focused on assessing the level of the mechanical vibrations generated by the system. For this purpose, three tri-axial accelerometers were mounted on the end of the manipulator arm and on the robot's frame. They measured the speed of mechanical vibrations in subsequent tests.

4 Conclusions

The presented computational model of the UAV with attached manipulator will be used to design dedicated manipulating device. The obtained results of the test-rigs during the flight present the complexity of the problem and give us practical information about the further construction and idea of attaching such a device to the robot. The most significant element in the process of the UAV integration with manipulator will be an application of the modified algorithm of the autonomous robot's control system in order to actively reduce the influence of the attached manipulator on UAV's dynamics.

References

Bar-Itzhack IY, Harman RR (2002) In-space calibration of a skewed gyro quadruplet. J Guidance, Control, Dyn 25(5):852–859

Doitsidis L, Valavanis KP, Tsourveloudis NC, Kontitsis M (2004) A framework for fuzzy logic based UAV navigation and control. In: Proceedings of the 2004 IEEE international conference on robotics and automation, New Orleans

Hald U, Hesselbaek M, Holmgaard J, Jedsen CH, Jakobsen S, Siegumfeldt M (2005) Autonomous helicopter—modeling and control. Aalbor University, Department of control engineering

Haug E (1989) Computer aided kinematics and dynamics of mechanical systems. Basic methods, vol 1. Allyn and Bacon, London

Heffley RK, Mnich MA (1988) Minimum-complexity helicopter simulation math model, NASA

Jang JS, Tomlin CJ (2003) Longitudinal stability augmentation system design for the DragonFly UAV using a single GPS receiver. Department of Aeronautics and Astronautics, Stanford University

Jensen D, Roberts M, Alifano J, Szumczyk T, Cheung W (2004) Development of PolyUAV: polytechnic university's unmanned aerial vehicle. Polytechnic University, AUVSI undergrad competition, Brooklyn

Kalman RE (1960) a new approach to linear filtering and prediction problems. Trans ASME–J Basic Eng (Ser D) 82:35–45

Kumar NS, Jann T (2004) Estimation of attitudes from a low-cost miniaturized inertial platform using Kalman Filter-based sensor fusion algorithm. Sadhana, Part 2, 29:217–235

Mao G, Drake S, Anderson BDO (2004) Design of a extended Kalman filter for UAV localization. School of electrical and information engineering, University of Sydney, Research school of information sciences and engineering, Australian National University

Merwe R, Wan EA (2004) Sigma-point Kalman filters for nonlinear estimation and sensor-fusion—applications to integrated navigation. OGI School of science and engineering, Oregon Health and Science University, Washington

Quigley M, Goodrich MA, Griffiths S, Eldredge A, Beard RW (2005) Target acquisition, localization, and surveillance using a fixed-wing mini-UAV and gimbaled camera. Computer Science Brigham Young University, USA

Sasiadek J, Zalewski J, Johnson R (2002) Kalman filter enhancement for UAV navigation. University of Central Florida Orlando, Carleton University, Canada

Tsai LW (1999) Robot analysis: the mechanics of serial and parallel manipulators. Wiley, New York

Welch G, Bishop G (2001) An introduction to the Kalman filter. Siggraph, Course 8, Los Angeles

Wood G, Kennedy D (2003) Simulating mechanical systems in simulink with simmechanics. Technical report, MathWorks, Natick

Zhang P, Gu J, Milios EE, Huynh P (2005) Navigation with IMU/GPS/Digital compass with unscented Kalman filter. In: Proceedings of the IEEE international conference on mechatronics and automation, Canada

The Kinematics Aspect of Robots Formation in Cooperation Tasks

Tomasz Buratowski, Józef Giergiel, Tadeusz Uhl and Andrzej Burghardt

Abstract The chapter presents a hierarchical control system for a formation of mobile robots used for large-size object transportation. The displacement and velocity values of the driving wheels in all robots were determined on the basis of inverse kinematics task. Proposed solution was simulated in an emulator of a n-robots' working environment and verified in an original control-measurement environment, which is capable of managing the operation of up to 256 mobile robots.

1 Introduction

The problem of large-size and great mass objects transportation is related both to the world of nature and technology. In the world of nature it has led to the series of solutions. Excellent example is how ants transport objects or the well-known problem of moving a table by 2 persons. Many works have raised the subject of using large-size transportation or of cooperation between systems or devices. These papers bring up, for example, the problem of pushing a box (Miyata et al. 2002), pushing an extremely heavy object with its own support in the form of wheels (Wang et al. 1994), conveyance of an object located over robots (Zielinski and Trojanek 2009), relocating an object by robots located next to it (Schenker et al. 2000) or transporting a beam by 2 robots (Humberstone and Smith 2000). The authors of mentioned papers most often use three-wheeled robots (Stilwell and Bay 1993), n-wheeled robots (Humberstone and Smith 2000), and track chain

T. Buratowski (✉) · J. Giergiel · T. Uhl
AGH University of Science and Technology, Cracow, Poland
e-mail: tburatow@agh.edu.pl

A. Burghardt
Rzeszów Technical University, Rzeszów, Poland

J. Sąsiadek (ed.), *Aerospace Robotics*, GeoPlanet: Earth and Planetary Sciences,
DOI: 10.1007/978-3-642-34020-8_10, © Springer-Verlag Berlin Heidelberg 2013

vehicles (Kosuge et al. 1998) as transportation robots. The tasks performed by systems or formations of transportation robots concern not only the relocation of a large mass, but also the cooperation between the manipulators placed on the vehicles (Khatib et al. 1996) or performing behavioral tasks in an unknown environment on the basis of signals from video cameras or scanners (Wang et al. 1994). The papers provide us with solutions based on considerable simplifications, e.g., robots and a relocated object are modelled with a point mass (Kosuge et al. 1998), as well as with solutions dealing with cooperation and mutual reaction between a robot and an object (Zielinski and Trojanek 2009). The construction presented in paper (Stilwell and Bay 1993) is an interesting solution that allows a robot formation to transport a large object connecting it with a centre of a two-wheeled mobile robot revolution using a link equipped with two rotary joints. Our chapter describes an inverse kinematics task for a system with four robots. The task makes use of a structure that is analogous to that of a fork-lift truck used in warehouses on a daily basis. In the chapter, the point of contact between the robot and the transported object is not located in the centre of the wheeled robot's rotation axis.

2 The Hierarchical System of Control Formation

The problem undertaken in this chapter is the transportation of large-scale objects by a group of robots (Fig. 1). The hierarchical system was adopted as the control architecture (Chmaj et al. 2007; Burghardt et al. 2011b) (Fig. 2).

The source of information on the defined trajectory of transported object motion is a superior control system or an operator, which provides the control system with information in the form of a vector $q_M = [x_M, y_M, \alpha]^T$. On its basis the inverse kinematics task is performed on the middle level of the control system. It generates the motion trajectory for all the robots participating in the task, provided there is

Fig. 1 The group of mobile robots

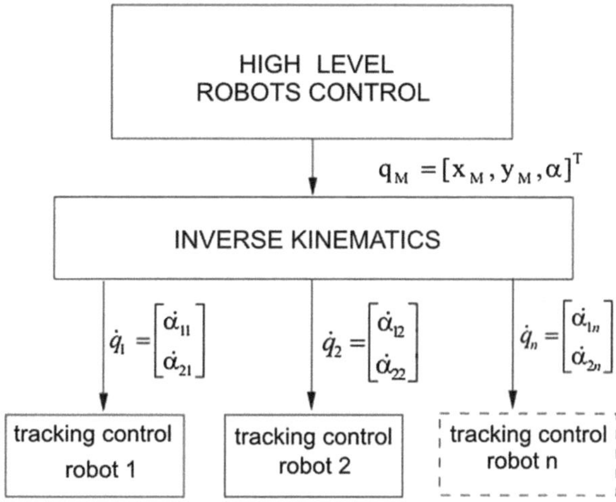

Fig. 2 The group of mobile robots

no skid in the contact point between the robot and the object. The lowest level of the control system, which is placed on the mobile robots, ensures the motion of the robots along the generated trajectory Burghardt et al. 2011a.

This chapter does not deal with the dynamics of the system, and the related reactions are treated as a disturbance to adaptation algorithms of the follow-up control. This is the lowest level of the hierarchical control system. The idea of using such algorithms was created for large-scale transportation tasks.

3 The Inverse Kinematics of the Robots Formation

Analyzing the inverse kinematics task of the n-robots system transporting the 0 body, we assume that the motion of all the robots will be performed on a horizontal surface. The chosen point which has coordinates x_M, y_M, which are connected with the transported rigid body, will move along the defined trajectory.

In the chapter, the robots are numbered from $i = 1$ to n. Let us assume that there is a motion joint, rotary kinematic pair of fifth class, in the points x_{Hi}, y_{Hi}. The joints are connected with the transported body and the robots.

To sum up, the presented inverse kinematics task involves determining the motion parameters for all the robots transporting the 0 object. In our case, we have mobile robots which move using two driving wheels; because of that, we should look for the displacement and velocity values of the driving wheels of all the robots $q_i = \begin{bmatrix} \alpha_{1i} & \alpha_{2i} \end{bmatrix}^T$ in order to ensure the M point's relocation in the 0 body along the defined trajectory.

Considering the given vector $\rho_{1Hi} = [x_{1Hi}, y_{1Hi}, 1]^T$ describing the coordinates of point H of nth robot in the local frame of reference x_1, y_1 related to the transported body, and using the transformation matrix $T_{1,0}$, we determine the coordinates of points x_{Hi}, y_{Hi} in the global frame of reference x, y in the following way:

$$\rho_{Hi} = T_{1,0}\rho_{1Hi} \tag{1}$$

where $\rho_{Hi} = [x_{Hi}, y_{Hi}, 1]^T$, and

$$T_{1,0} = \begin{bmatrix} \cos(\alpha) & -\sin(\alpha) & x_M \\ \sin(\alpha) & \cos(\alpha) & y_M \\ 0 & 0 & 1 \end{bmatrix}$$

The decomposition of the velocity vectors for points A_i suggests that their projections into the axes of the basic frame of reference x, y fulfill the following equation:

$$\dot{y}_{Ai} = \dot{x}_{Ai} tg\beta_i \tag{2}$$

The β_i are the angles of rotation of robots' frames on a planar surface. It means that the non-holonomic restraints are imposed on the velocity vectors of the A_i points.

On the basis of the system geometry (Fig. 3), the following dependencies exist for the first and fourth robot ($i = 1$, $i = 4$)

$$x_{Hi} = x_{Ai} + l_3\cos(\beta_i), y_{Hi} = y_{Ai} + l_3\sin(\beta_i) \tag{3}$$

and the following for the robots two and three ($i = 2$, $i = 3$)

$$x_{Hi} = x_{Ai} - l_3\cos(\beta_i), y_{Hi} = y_{Ai} - l_3\sin(\beta_i) \tag{4}$$

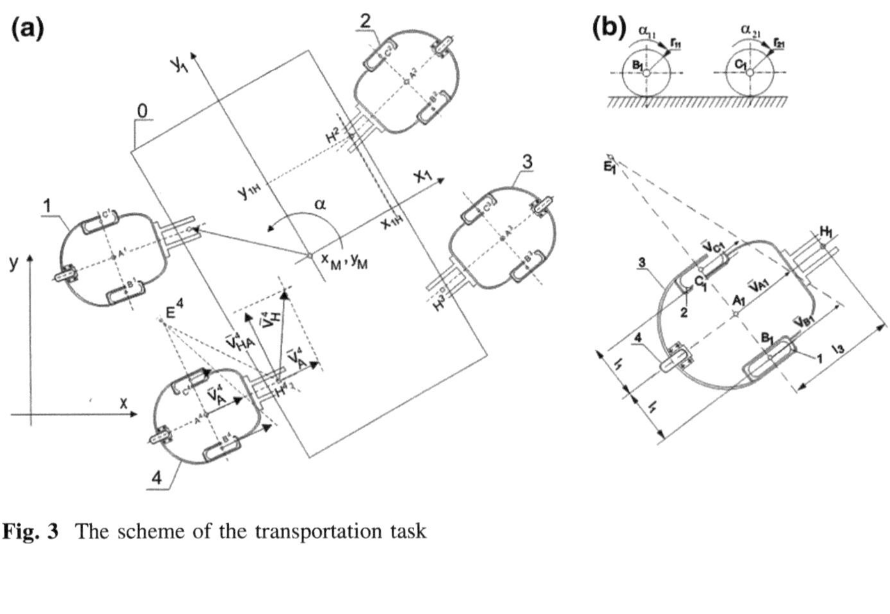

Fig. 3 The scheme of the transportation task

The further discussion deals with Eq. (3), which describes location of the *H* points of the robots 1 and 4. In order to determine the inverse kinematics task for robots 2 and 3, we need to act by analogy with steps below, using Eq. (4).

Differentiating the dependence (3) over time, we obtain:

$$\dot{x}_{Hi} = \dot{x}_{Ai} - l_3\dot{\beta}_i \sin(\beta_i), \dot{y}_{Hi} = \dot{y}_{Ai} + l_3\dot{\beta}_i \cos(\beta_i) \tag{5}$$

On the basis of the decomposition of *H* point velocity vectors, we can write:

$$\bar{v}_{Hi} = \bar{v}_{Ai} + \bar{v}_{HAi} \tag{6}$$

As the velocity vector of point A_i is perpendicular to the relative velocity vector v_{HAi} then

$$(v_{Hi})^2 = (v_{Ai})^2 + (v_{HAi})^2 \tag{7}$$

Let us write the relative velocity as: $v_{HAi} = \dot{\beta}_i l_3$, $(v_{Ai})^2 = (\dot{x}_{Ai})^2 + (\dot{y}_{Ai})^2$, and $(v_{Ai})^2 = (\dot{x}_{Ai})^2 + (\dot{y}_{Ai})^2$; then, on the basis of dependencies (5) and (7), we obtain:

$$\dot{\beta}_i = \frac{-\dot{x}_{Hi}\sin(\beta_i) + \dot{y}_{Hi}\cos(\beta_i)}{l_3} \tag{8}$$

Equations (5) and (1) allow us to write:

$$\begin{aligned} \dot{x}_{Ai} &= \dot{x}_{Mi} - \dot{x}_{1Hi}\dot{\alpha}\sin(\alpha) - \dot{y}_{1Hi}\dot{\alpha}\cos(\alpha) + l_3\dot{\beta}_i \sin(\beta_i) \\ \dot{y}_{Ai} &= \dot{y}_{Mi} + \dot{x}_{1HI}\dot{\alpha}\cos(\alpha) - \dot{y}_{1Hi}\dot{\alpha}\cos(\alpha) - l_3\dot{\beta}_i \cos(\beta_i) \end{aligned} \tag{9}$$

The description in (8) and (9) form presents the motion parameters of the bodies numbered from 1 to 4, i.e., the robots. The further discussion determines dependencies between the vector $\left[\dot{x}_{Ai}, \dot{y}_{Ai}, \dot{\beta}_i\right]^T$ describing the velocity of an nth robot and velocities of the driving wheels.

It is assumed that the values of the linear velocities of the B_i, C_i points are matched in such a way that $V_{Bi} > V_{Ci}$ (robots 1, 4); because of this, the frames of the robots move on a planar surface. The points E_i, shown in Fig. 3, are the instantaneous center of rotation of nth robot's frame. Equations describing the velocities of points B_i and C_i have been written in the following form:

$$\begin{aligned} \bar{v}_{Ci} &= \bar{v}_{Ai} + \bar{v}_{CAi}, \quad v_{Ci} = v_{Ai} - v_{CAi}, \\ \bar{v}_{Bi} &= \bar{v}_{Ai} + \bar{v}_{BAi}, \quad v_{Bi} = v_{Ai} + v_{BAi}, \end{aligned} \tag{10}$$

in addition, we assume that:

$$v_{BAi} = v_{CAi} = \dot{\beta}_i \cdot l_1 \tag{11}$$

Assuming that the driving wheels cooperate with surface without skid between wheel and ground, we write:

$$v_{Bi} = \dot{\alpha}_{1i} \cdot r_1, v_{Ci} = \dot{\alpha}_{2i} \cdot r_2 \tag{12}$$

where $r_1 = r_2 = r$ are the radii of the robots' driving wheels.

On the basis of Eqs. (11) and (12), we obtain:

$$\begin{aligned} \dot{\alpha}_{1i} &= r^{-1}\left(\sqrt{(\dot{x}_{Ai})^2 + (\dot{y}_{Ai})^2} + l_1\dot{\beta}_i\right) \\ \dot{\alpha}_{2i} &= r^{-1}\left(\sqrt{(\dot{x}_{Ai})^2 + (\dot{y}_{Ai})^2} - l_1\dot{\beta}_i\right) \end{aligned} \tag{13}$$

To sum up, kinematic dependencies (8), (9) and (13) allow us to determine the inverse kinematics task for the robots formation. In other words, we are able to determine angular velocities of the driving wheels of all the robots participating in the transportation task for the defined 0 body motion, in the form of vector of the parameters $q_M = [x_M, y_M, \alpha]^T$.

4 Simulation and Verification

The simulation of the proposed solution was performed in the Matlab/Simulink software, using mobile n-robots' working environment, built by the authors. For a defined transported object, the M point motion trajectory, presented in Fig. 4a, was adopted The M point linear velocity and the *0* body angular velocity (Fig. 4b) were also assumed. Figure 4d presents the visualization of the object transportation task,

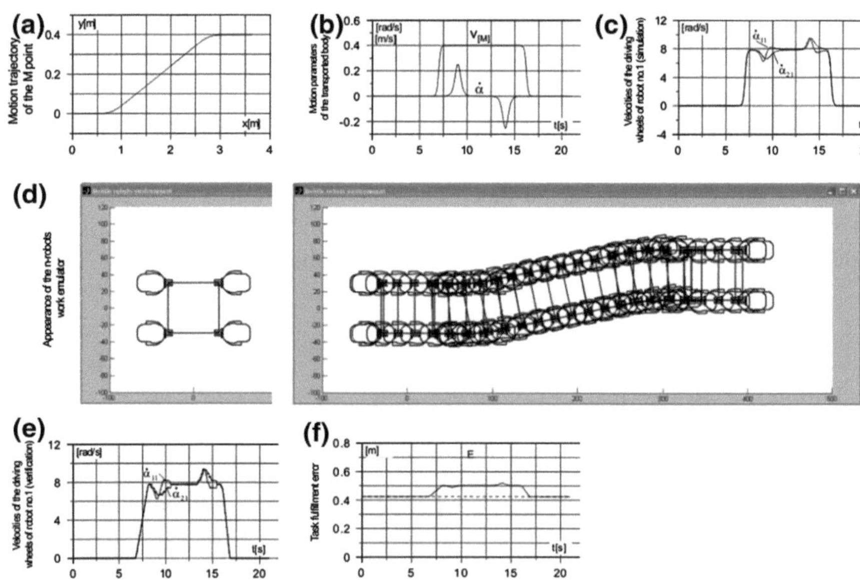

Fig. 4 The results of simulation and verification

and Fig. 4c presents values of determined velocities of the driving wheels for the chosen mobile robot.

In order to qualitatively compare the behavior of a formation shape during verification, when various types of trajectory were tested, we introduced a quality coefficient in the following form:

$$E = \sum_{i=1}^{n} \sqrt{(x_M - x_{Hi})^2 + (y_M - y_{Hi})^2} \quad \text{where } n = 4 \tag{14}$$

Verification of the proposed formation control algorithm was obtained in the control-measurement environment. In order to record signals from an actual robots, we developed software in the form of Matlab/Simulink toolboxes (Burghardt 2008). This solution allows us to monitor the work of a group of mobile robots. The toolboxes allow recording of: motion parameters, data from distance sensors, battery voltage and value of current in wheel driving motors. The other function of these tools is drive control. The appearance of the configuration window of a subsystem created in Matlab/Simulink environment is shown in Fig. 4d. During the experimental research, the object was transported on the trajectory analogous to the simulated one. The simulation results (Fig. 4e, f) validate the initial assumptions of the project, and the proposed control algorithm allows us to control the motion of the defined formation with hierarchical control system implementation. During verification using mobile wheeled robots, we obtained results that are analogous to those from the simulation.

The error values of the formation shape during the mobile robots system actual motion are greater, than those received from simulation, by about 15 %. This difference is a result of modeling errors and existence of disturbances that were not modeled or simplified in the simulations.

5 Conclusions

The adopted simplifying assumptions allowed us to obtain a description of kinematics of the robots formation. Simulations and verification have corroborated correctness of obtained mathematical description of the motion. Other conclusions derived from the verification research, direct further research works to a more comprehensive modeling of the contact between robots and the transported object and to modeling dynamic reactions acting during the motion. Moreover the problem of precise navigation should be taken into consideration in order to remove influence of the systematic error (Burghardt et al. 2011b), connected with the real model of motion which takes into account skid and uncertainty of the measured distances between significant points of the robot's frame. The idea of using precise navigation generally is connected with skid reduction with the use of probabilistic algorithms often utilizing Kalman filter (Banaszkiewicz et al. 2010).

The proposed approach to the inverse kinematics task for a group of robots in formation give us the possibility to jointly transport an object. Additionally to the received collective behavior it is possible to trace the kinematic parameters for all the robots in formation.

References

Banaszkiewicz M, Buratowski T, Dabrowski B, Uhl T (2010) The precise odometry navigation for the group of robots. Schedae Informaticae 19:99–111

Burghardt A (2008) Proposal for a rapid prototyping environment for algorithms intended for autonomous mobile robot control. Mech Mech Eng, Technical University of Lodz 1:5–16

Burghardt A, Buratowski T, Gergiel J (2011a) Control of a robots' formation in unknown surroundings environment. In: Dynamical systems: nonlinear dynamics and control, Wydawnictwo Politechniki Łódzkiej, Łódź pp 95–102

Burghardt A, Buratowski T, Giergiel J, Uhl T (2011b) The autonomous group of robots precise navigation. Pol J Environ Stud 20(5A):35–40

Chmaj G, Buratowski T, Uhl T (2007) The project of an autonomous robot capable to cooperate in a group. In: Kozlowski K (ed) Robot motion and control 2007 (lecture notes in control and information sciences). Springer, London 391–399

Humberstone CK, Smith KB (2000) Object transport using multiple mobile robots with non-compliant end effectors. In: Distributed autonomous robotics systems. Springer, Tokyo 4:417–426

Khatib O, Yokoi K, Chang K, Ruspini D, Holmberg R, Casal A (1996) Vehicle/arm coordination and multiple mobile manipulator decentralized cooperation. IEEE/RSJ international conference on intelligent robots and systems

Kosuge K, Osumi T, Sato M, Chiba K, Takeo K (1998) Transportation of a single object by two decentralized-controlled non-holonomic mobile robots. IEEE international conference on robotics and automation, Belgium, vol 4, pp 2989–2994

Miyata N, Ota J, Arai T, Asama H (2002) Cooperative transport by multiple mobile robots in unknown static environments associated with real-time task assignment. IEEE Trans Robot Autom 18(5):769–780

Schenker PS, Huntsberger TL, Pirjanian P, Trebi-Ollennu A, Das H, Joshi S, Aghaz-Arian H, Ganino AJ, Kennedy BA, Garrett MS (2000) Robot work crews for planetary outposts: close cooperation and coordination of multiple mobile robots. In: Proceedings SPIE symposium on sensor fusion and decentralized control in robotic system III, Boston

Stilwell DJ, Bay JS (1993) Towards the development of a material transport system using swarms of ant-like robots. IEEE/RSJ international conference on intelligent robots and systems, Atlanta

Wang Z, Nakano E, Matsukawa T (1994) Cooperating multiple behavior-based robots for object manipulation. IEEE/RSJ international conference on intelligent robots and systems

Zielinski C, Trojanek P (2009) Stigmergic cooperation of autonomous robots. J Mech Mach Theory 44:656–670

Ultra-Light Planetary Manipulator: Study and Development

Jerzy Grygorczuk, Bartosz Kędziora, Marta Tokarz, Karol Seweryn, Marek Banaszkiewicz, Marcin Dobrowolski, Paweł Łyszczek, Tomasz Rybus, Michał Sidz, Konrad Skup and Roman Wawrzaszek

Abstract The demand for the use of Planetary Manipulators is quite obvious in space exploration and in situ research. It can also be predicted that its role will expand in upcoming decades given the planned intensification of space exploration. This chapter presents the newly developed 3dof Ultra-Light Planetary Manipulator (ULPM) designated for extended servicing of exploration tools (e.g., penetrators, small rovers, etc.), or scientific instruments and sensors in planetary missions, where Mars and Moon are the mostly foreseen destinations. It combines new ideas and earlier achievements, both of which had influence on the concept and would demonstrate the technology. In consequence, a laboratory model device was successively developed. Two leading constraints determined the design: very low mass of the unit and long extension range of deployment of the servicing instruments. For those reasons, the utilization of the tubular boom mechanism was preferred. Integrated Ultra-Light Planetary Manipulator was tested in laboratory conditions in order to prove its operational functionality and performance. Particularly, it was examined what is the safety deployment distance for certain loads with acceptable stability of the flexible system structure. This topic was also one of the most important issues for performed analysis and manipulator's dynamics simulations.

1 Concept

Since 1996 SRC PAS has been involved in the development of various low speed penetrators, including mole type ones which are driven by hammering (Grygorczuk et al. 2011; Seweryn et al. 2012). Penetrators of this kind, practically without

J. Grygorczuk (✉) · B. Kędziora · M. Tokarz · K. Seweryn · M. Banaszkiewicz · M. Dobrowolski · P. Łyszczek · TomaszRybus · M. Sidz · K. Skup · R. Wawrzaszek
Space Research Centre of the Polish Academy of Sciences,
Bartycka 18A, 00-716 Warsaw, Poland
e-mail: jurekgry@cbk.waw.pl

J. Sąsiadek (ed.), *Aerospace Robotics*, GeoPlanet: Earth and Planetary Sciences,
DOI: 10.1007/978-3-642-34020-8_11,

exception, have to be serviced by another dedicated subsystem (mostly manipulators) located on-board the lander. A discussion on what a desired servicing manipulator has to provide, besides mandatory for space devices lock and release functions, is listed below:

- Unrestricted selection of the operational site, not limited by the maneuverability (degrees of freedom) of the manipulator. That means that 3dof system is most preferable: extension, lifting up and down, rotational joint in horizontal plane.
- Sufficient stability. This is a real problem for a manipulator which has to be ultra-light and requires relatively long extension range (3 m) in comparison to the foreseen small envelope for stowed configuration, and, on the other hand, has to provide accurate and durable operation. Three-meter extension range is taken into account as in Martian descending technics the foil airbag surroundings the lander is used. After remaining seated, the foil is pierced by a pyro and can spread out to up to 2.5 m (ExoMars mission data) from the lander. Another benefit of the long extension is related with not being shadowed and with possibly performing better measurements.
- Retracting. Depending on the intended applications, the manipulator in principle could deploy the instrument and remain in this position. For better universality, e.g., not obstructing the instrument's measurements by its presence or being ready for subsequent action, the retracting feature is demanded.
- Rotational joint in horizontal plane. This could certainly be an option. One user might require it and another might have to give it up, e.g., in terms of the mass limits. In modular design, any choice would not constitute a problem. It was decided to implement the rotational joint in the prototype model.
- End-effector performance. In principle, at the manipulator's free end there can be a simple gripper for holding the tool (penetrator, small rover, special sensors, etc.) or a specialized end-effector with additional 2–3dof, if, e.g., perpendicularity of the penetrator to the ground would be indispensable. In the development of a new type of ULPM, the end-effector can be skipped entirely in the prototype model. Its presence and properties will be always dependent on a specific payload and would have negligible impact on the baseline tests.
- Special functions, e.g., support for initial insertion of the penetrator, equipped with depth sensors, etc. As in the previous point this is not an important issue for the prototype model.

Summarizing the above constraints, it is obvious that the main challenges concerning this Ultra-Light Planetary Manipulator are the relatively large dimensions, the necessary to handle distance between lander upper platform and the planetary surface, the very restricted mass budget, the environmental conditions, and the required operational flexibility. Another critical parameter for the manipulator's design is its compactness (in stowed configuration) in comparison with its extension range (about 3 m). The acceptable mass for the manipulator is estimated at 2.0–2.5 kg. The value is based on the experience from the previous space missions, where even the given mass had to be greatly decreased later on.

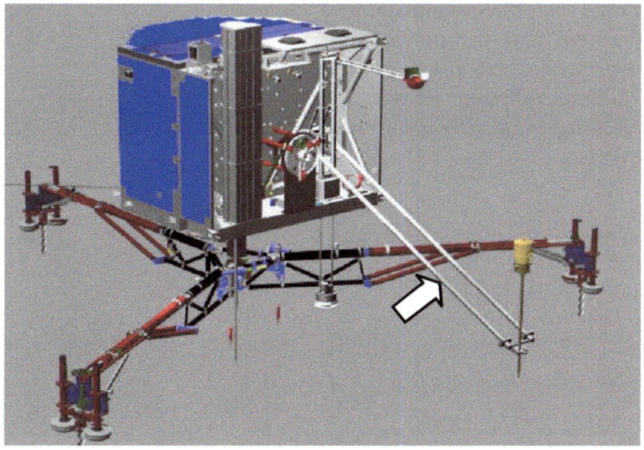

Fig. 1 MUPUS—penetrator deployed by the tubular boom manipulator (marked by *arrow*) from the Philae lander

Off-the-shelf solutions are not available and a new design of these mechanisms has to take into account common principles as well as new approaches (Paulikas et al. 2007, Smith et al. 2009). SRC PAS has a lot of experience in developing dedicated space robotic mechanisms, also with unconventional solutions. A good example is a light-weight manipulator that has been developed for the MUPUS penetrator deployment (Grygorczuk and Banaszkiewicz 1998; Grygorczuk et al. 2007), for the Rosetta mission 10 years ago that is now approaching the 67P /Churyumov-Gerasimenko comet (Spohn et al. 2006). In that manipulator, an original tubular boom system was chosen for the translation motion. After careful analysis and simulations, the manipulator passed all qualification and acceptance tests (TRL8), and was integrated with the lander's balcony (Fig. 1).

Tubular boom (TB)—a rounded spring metallic tape—is a light-weight material (40 g/m for ø15 × 0.1 mm tape, or 80 g/m for ø25 × 0.15 mm tape) and combined with very convenient storage of the TB, gives promising areas of application. Tubular boom is relatively stiff due to its round shape and can be stowed by simply rolling it up on the reel and it can be released for translational motion by unwinding it from the reel, without special concern as to the length. But they also have a week point: unacceptably low torsional stiffness of their open profile, which has to be improved. One of the methods is to array a parallel boom system giving much stiffer solution both for rolling and bending torques. Parallel boom system was used with good results for the MUPUS manipulator and it is shown in Fig. 2.

When considering a new Ultra-Light Planetary Manipulator, one of the ideas was to build upon this valuable solution as much as possible. However, the concept had to be revised and developed for very specific and more challenging requirements. MUPUS manipulator had only 1 m extension distance and did not have 3dof. The most crucial for the ULPM concept was the invention of a guy support mechanism, which, besides running the second degree of freedom (up and down elevation), would provide intrinsic improvement of applying loads.

Summarizing the above, a principle of operation is shown in Figs. 3, 4 and 5. Figure 3 shows the ULPM in stowed configuration. A tool is represented by a rod

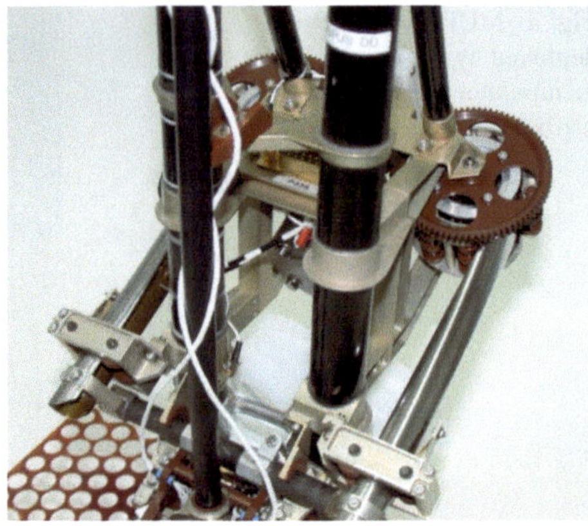

Fig. 2 MUPUS manipulator (stowed). Two parallel booms were arrayed on distance of 120 mm, rolled up on two reels and guided by the front rollers

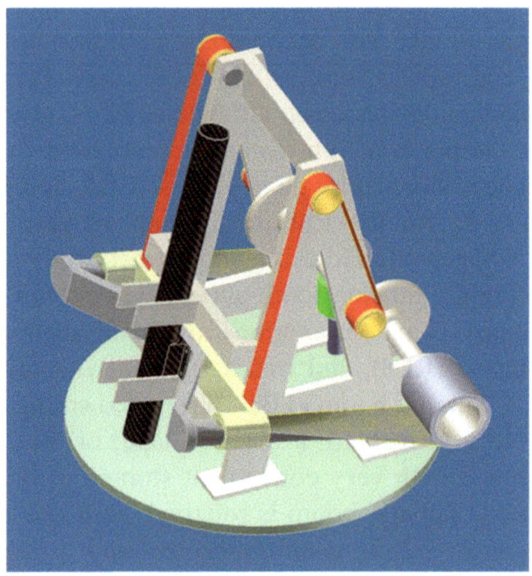

Fig. 3 Manipulator stowed

(e.g., like a mole penetrator). The 2dof manipulator (translational extension and rotational tilting) is fixed to the rotational platform which adds the third DOF.

Figure 4 shows the ULPM conceptual system during operation. The tubes are tilted up to minimise the risk of hiting obstacles. The guy strips are fixed to the slider which will move forward using the TBs as a guide. A special route of the strip will provide for the position of the slider to be set up in the middle of the extended tubes. At any given time the guy strips' length will determine the tilting position. The arrows show the forces directions generated in tubular booms and guy strips by the load, under gravity conditions.

Figure 5 shows the full length extension and lowering of the payload to the ground. The ULPM can be retracted to the stowed configuration (by rolling up the tubular booms) and used again to deploy and service another instrument.

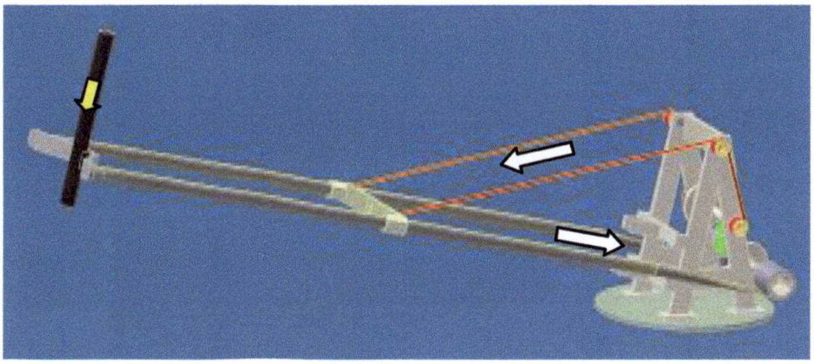

Fig. 4 During operation. *Arrows* demostrate that the guy strips are axially tensioned and the tubular booms are axially compressed in the indicated area

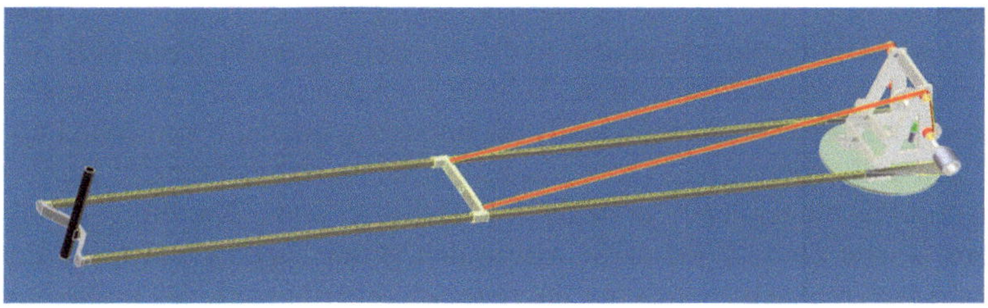

Fig. 5 Maximum length of deployment

2 Design Features

In general, the scope of this chapter is limited to the most important design features. One of the key issues was the choice of a proper tubular boom type, configuration of two parallel booms and their drive. From available TB profiles, one was chosen that is stronger than the profile which was applied in MUPUS manipulator. It is made of stainless steel 0.15 mm thick tape whose regular width is 90 mm. The tape is rounded to circular 25 mm diameter with about 10 mm overlap. In this design the tape's width was reduced to 70 mm and instead of overlapping, "C" profile was acquired, which in this specific application offers the following advantages:

- Shorter distance of tape profile transformation needed for compactness of the design. The TB profile on the reel is flat and rises up to the full circular profile on a certain distance. For 90 mm width, the profile transformation distance is over 0.5 m, and for reduced width it is accordingly 1/3 less (nonlinear dependence).
- Mass reduction with acceptable decrease of stiffness.
- Narrower tape corresponds with a more compact design.

Fig. 6 A new translation motion drive which is based on meshing of the TB's holes with the pins toothed wheel

The next design feature is related to the force acting on the boom (Fig. 4) and to the precision of translation motion. Tubular booms are axially compressed in the root area. This force is not sufficiently supported by the flat boom rolled up on the reel. The axially compressed force causes the tight winding boom to get loose and the support has the tendency to be flexible and inaccurate. Therefore, to solve that problem, a new kind of TB drive with much stiffer support had to be invented. Finally, it was realized by making the regularly distributed holes in the middle of the boom. The holes are meshed with pins toothed wheel, and linear forward or backward translation motion is achieved by the rotation of the wheel. The support is stiff and the motion is precise, and most importantly—with little backlash. The presented solution is shown in Fig. 6.

An important role in this design is played by the guy strips which provide:

- One degree of freedom through motion in vertical plane by tilting up or down the tubular booms.
- Improving the load capacity and stiffening the extension system. The guy strips' strength and elasticity have significant influence on that.
- Tubular booms are always supported in their middle independently of their ends' positions. The guy strips maintain the cart in proper position through specially designed set of rollers and fixations.

In Fig. 7, the guy strips subsystem design is shown, i.e., how they are routed and where they are fixed. Additionally, the subsystem is equipped with telescopic mechanism which lifts upper fixation of the guy strips. In stowed configuration, the ULPM has 350 mm of height. Thanks to the mechanism, it gains additional 250 mm. It results in a larger support angle, a more optimal solution for such a long, end-loaded boom.

The cart shown in Fig. 8 has a set of rollers designated for two different drives. One of them is used for the guy support system. The other one has a specially designed profile and serves to guide the cart along the tubular booms.

Another goal was to design a light-weight rotational joint mechanism, which is foreseen as an optional equipment. It gives one additional dof, which makes the

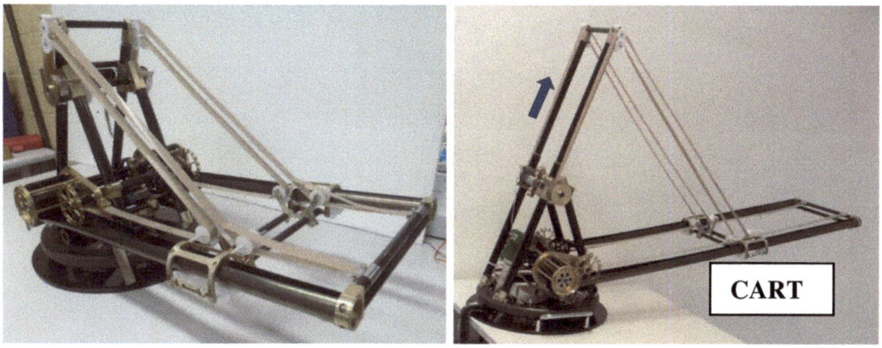

Fig. 7 Guy strips subsystem. Within released telescopic mechanism (*left*). Operational configuration, the mechanism extends the *upper* fixation of the guy strips (*right*)

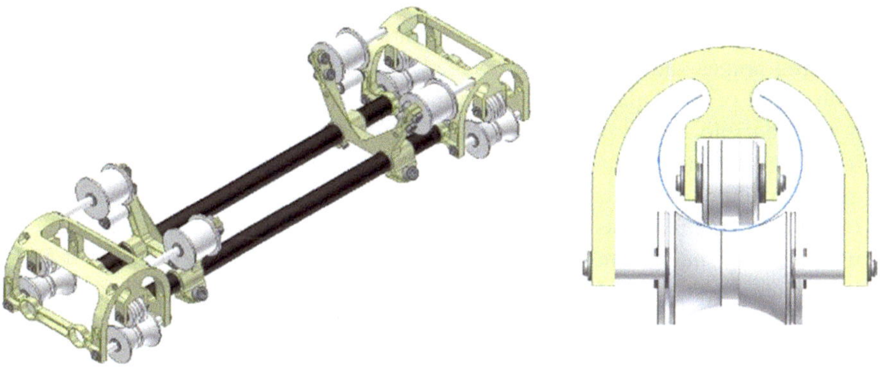

Fig. 8 The cart with a set of rollers (*left*). Profile of the rollers used for guiding with tubular boom (*right*)

ULPM more of a universal device. Forgoing this module, e.g., if there are weight restrictions, causes the ULPM to become lighter by about 600 g. As shown in Fig. 9 the rotational joint mechanism consists of CFRP C shaped ring and three sets of rollers. Thanks to carbon fiber, the base ring has high mechanical strength as well as stiffness and is also very light. Similarly to the tubular boom drive mechanism, a toothed wheel, with specially designed pins cooperating with slots drilled in the base ring, was applied as propulsion.

3 Conducting Tests

In this chapter, preliminary tests of integrated prototype of the Ultra-Light Planetary Manipulator, which have been performed at SRC PAS laboratory, are presented. During the tests the following performance aspects of the manipulator were checked and determined:

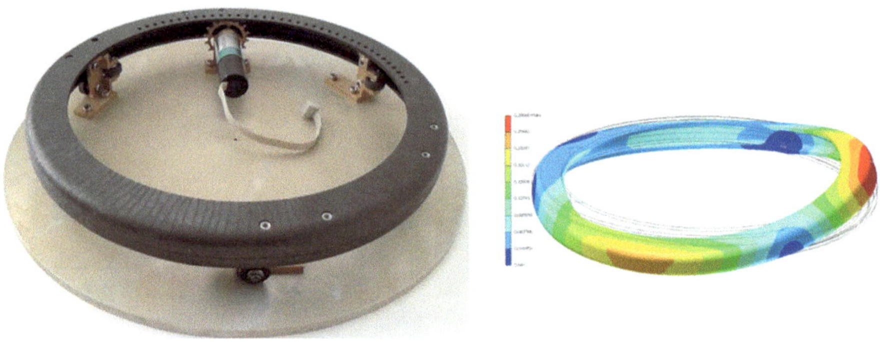

Fig. 9 Rotational joint mechanism fixed on the baseplate (*left*). FEM calculations of the CFRP C shaped ring (*right*). The ring can withstand 300 N loads

- Kinematics and range of each degree of freedom. The tests confirmed that all drives worked smoothly, power consumption was about 3 W, and that all required ranges and speeds were reached. Maximum extension length was 3.0 m with speed of 1 m/min. It was determined that reaching slightly longer distance is also feasible. In the prototype the tilt angle was between +30° and −15° and the joint rotation angle was between ±60°, but those angles can be raised in the design of the next model. Figure 10 shows the ULMP fully extended.
- In the next stage of tests, the ULMP was loaded with a mass of 1.4 kg fixed on its free end. The system passed all tests. Figure 11 shows the loaded ULMP.
- Preliminary test concerning stability was done with 1.4 kg load extended to 2.2 m. The free end was additionally deflected by about 15 mm and released. The flexible system quickly (within few seconds) gets back to the state of equilibrium. Several design features have influenced it. One of them is the cart (coupled with guy strips) which during tubular booms deflections must slide

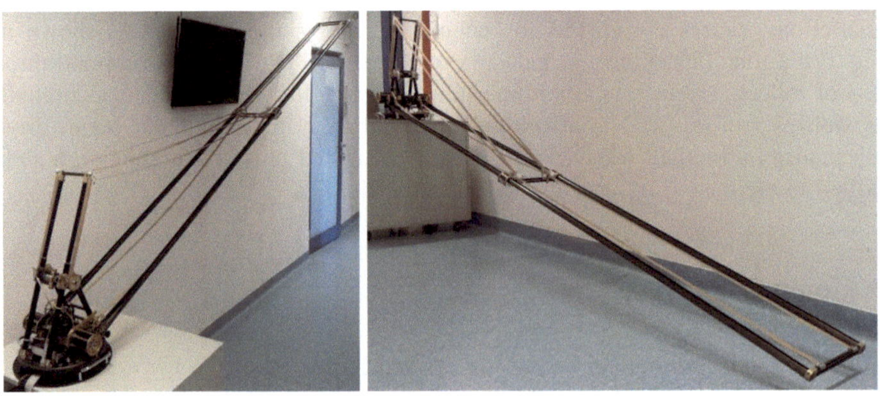

Fig. 10 ULMP during tests. It is fully extended (3.0 m) and tilted up (*left*) or tilted down (*right*). Difference in height of the free end between extreme positions is about 2.5 m

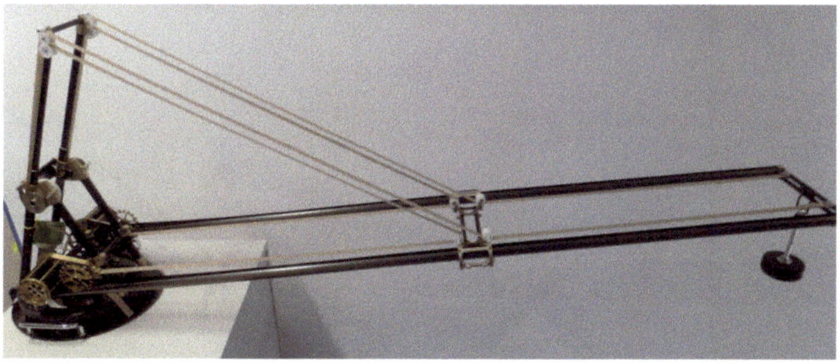

Fig. 11 Prototype of the ULMP with 1.4 kg load during tests

forwards and backwards on the booms resulting in dissipation of energy. Measured eigenfrequency for 1.4 kg load and 2.2 m extension was about 3 Hz.
- The last test concerning stability was conducted by twisting the free end beam by 45°. In normal circumstances it should not happen, but the system rapidly regained the state of equilibrium.

There are plans to perform tests with 5–10 kg loads with simulated Martian and Lunar gravity. A special set-up is under development.

4 Simulations and Analysis

On the simulation level, the ultra-light manipulator was described as a serial manipulator which consists of three kinematic pairs: two rotational and one translational. Kinematic scheme of the manipulator is presented in Fig. 12.

In principle, the manipulator includes flexible shape and, therefore, the dynamics of such system is described by the partial differentia equation (PDE) which can be simplified using assumed mode method—the ordinary differential equation (ODE) (Junkins and Kim 1993).

$$\begin{aligned} &[M_{rr}(q_r, q_e)]\{\ddot{q}_r\} + [M_{re}(q_r, q_e)]\{\ddot{q}_e\} + \{h_r(q_r, q_e, \dot{q}_r, \dot{q}_e)\} = \{\tau\}, \\ &[M_{er}(q_r, q_e)]\{\ddot{q}_r\} + [M_{ee}(q_r, q_e)]\{\ddot{q}_e\} + [K_e]\{q_e\} + \{h_e(q_r, q_e, \dot{q}_r, \dot{q}_e)\} = \{0\} \end{aligned} \tag{1}$$

where M is mass matrix, q_r and q_e are rigid and elastic elements of state vector, respectively, K is stiffness matrix and h_r is the vector which includes Coriolis term. The external torque is described by τ. Nevertheless, for determination of the level of torque appearing during typical manipulator's operation, the standard multi-body approach (Tsai 1999) for rigid bodies was taken into account and the elastic term in Eq. 1 was neglected.

To solve these equations, the simulation tool was developed at the SRC PAS based on SimMechanics software (part of the MATLAB Simulink) (Seweryn et al.

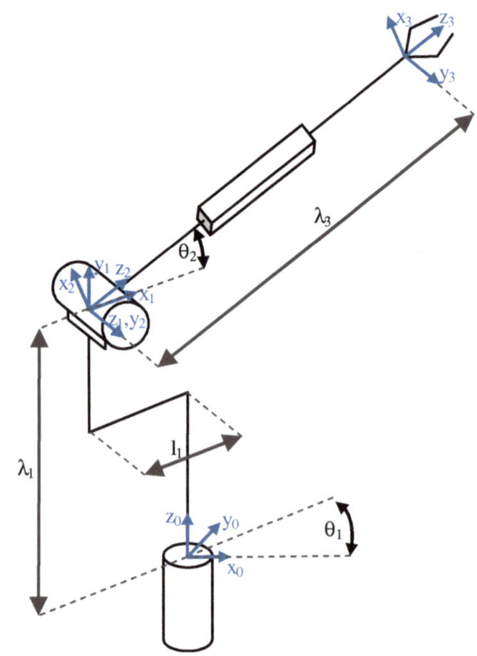

Fig. 12 Kinematic scheme of the manipulator

2011). The use of this platform allows incorporating in one environment all the elements essential in numerical simulations of the manipulator's dynamics (Wood and Kennedy 2003). The detailed model of the first rotational joint of the manipulator is presented in Fig. 13.

To compute torques and forces appearing in the joint, the following boundary conditions were set.

Link number	Start value	End value
Angle θ_1 (deg)	−23	23
Angle θ_2 (deg)	28	−6
Length λ_3 (m)	0,1	3

The simulation time was set as follows: 75 s for acceleration, 150 s for movement with constant speed and 75 s for breaking. The simulations were conducted for moon gravity conditions. The reference joint trajectory is presented in Fig. 14.

For such a joint movement, the torques obtained are shown in Fig. 15.

The results clearly show that the highest torque appears on the second joint, which has to balance the gravity force. It was one of the reasons that this torque was obtained in real solution by adding the telescoping mechanism. In the

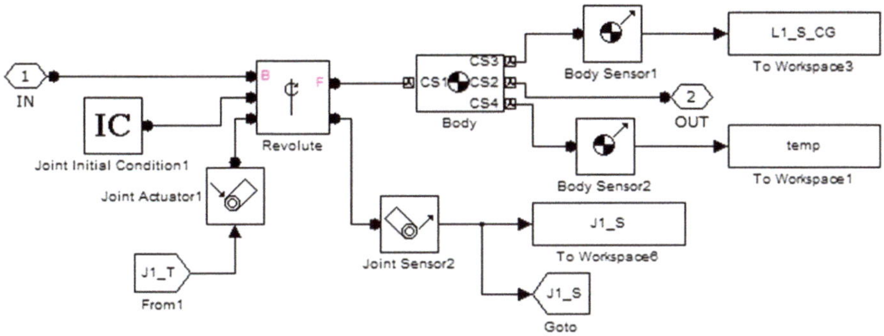

Fig. 13 Simulink model of manipulator's first joint

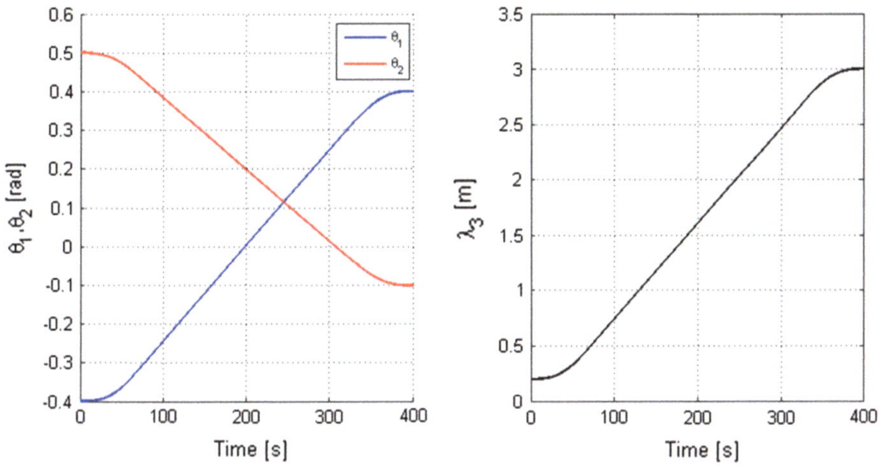

Fig. 14 Planned joint movement

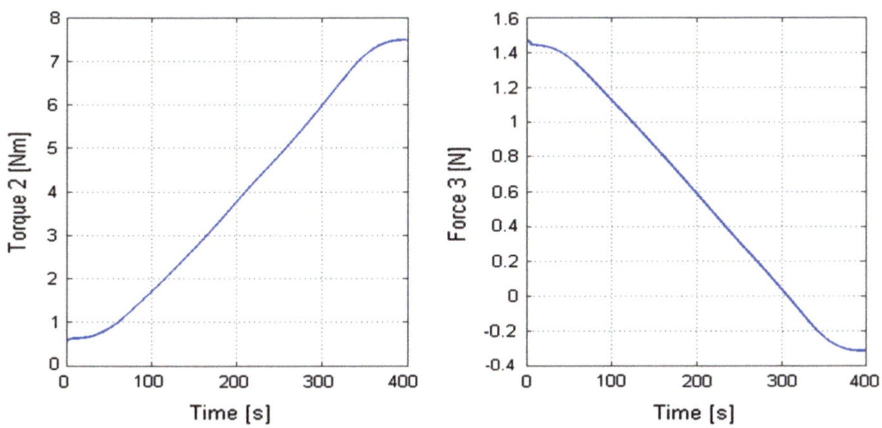

Fig. 15 The torque/force for the second and third joint

presented case the joint angle θ_2 crosses the zero position and, therefore, the gravity force causes the change of sign of the force 3 (Fig. 15). This effect can be a reason for the potential problem during control of the trajectory due to the backlash effect.

5 Conclusions

The integrated prototype of the newly developed Ultra-Light Planetary manipulator confirmed that the device is genuinely light, and the performed tests proved that ULPM is stable and can provide secure payload travel on a distance of 3 m.

The attained properties for the ultra-light planetary manipulator are given below.

Envelope (stowed) L × W × H	350 × 350 × 400 mm
Mass with (without) rotational joint	2.7 kg (2.1 kg)
Deployment length	3.0 m
Deployment speed	0.25–1.0 m/min
Tool mass with end-effector	Depends on extension and gravity: 1.4–10 kg
Power consumption	2.0–3.0 W
Tilting angle	+30°−−15°
Horizontal (azimuth) turn	±60°

Acknowledgments The chapter was supported by the NCN project no N N514 234537.

References

Grygorczuk J, Banaszkiewicz M (1998) A concept study of the deployment device for the MUPUS penetrator on the Rosetta lander. (Rosetta –MUPUS, internal report)

Grygorczuk J, Banaszkiewicz M, Seweryn K, Spohn T (2007) MUPUS insertion device for the Rosetta mission. J Telecommun Inf Technol 1:50–53

Grygorczuk J, Banaszkiewicz M, Cichocki A, Ciesielska M, Dobrowolski M, Kędziora B, Krsowski J, Kuciński T, Marczewski M, Morawski M, Rickman H, Rybus T, Seweryn K, Skocki K, Spohn T, Szewczyk T, Wawrzaszek R, Wiśniewski Ł (2011) Advanced penetrators and hammering sampling devices for planetary body exploration. In: Proceedings of the 11th symposium on advanced space technologies in robotics and automation, ESA/ESTEC, Noordwijk

Junkins J, Kim Y (1993) Introduction to dynamics and control of flexible structures. AIAA, Washington

Paulikas GA, Pieters CM, Banerdt WB, Burch JL, Chaikin A, Cohen BA, Duke M, England AW, Hiesinger H, Hinners NW, Howard AM, Lawrence DJ, Lester DF, Lucey PG, Stern SA, Tompkins S, Valero FPJ, Valley JW, Walker CD, Woolf NJ (2007) The scientific context for the exploration of the Moon. National Research Council, Washington

Seweryn K, Banaszkiewicz M, Maediger B, Rybus T, Sommer J (2011) Dynamics of space robotic arm during interactions with non cooperative objects. In: Proceedings of 11th symposium on advanced space technologies in robotics and automation (ASTRA)

Seweryn K, Grygorczuk J, Wawrzaszek R, Banaszkiewicz M, Rybus T, Wiśniewski Ł (2012) Low velocity penetrators (LVP) driven by hammering action—definition of the principle of operation based on numerical models and experimental tests. Submitted to mechanism and machine theory

Smith A, Crawford IA, Gowen RA, Ball AJ, Barber SJ, Church P, Coates AJ, Gao Y, Griffiths AD, Hagermann A, Joy KH, Phipps A, Pike WT, Scott R, Sheridan S, Sweeting M, Talboys D, Tong V, Wells N, Biele J, Chela-Flores J, Dabrowski B, Flannagan J, Grande M, Grygorczuk J, Kargl G, Khavroshkin OB, Klingelhoefer G, Knapmeyer M, Marczewski W, McKenna-Lawlor S, Richter L, Rothery DA, Seweryn K, Ulamec S, Wawrzaszek R, Wieczorek M, Wright IP, Sims M (2009) Lunar EX—a proposal to cosmic vision. Exp Astron J. doi 10.1007/s10686-008-9109-6A

Spohn T, Seiferlin K, Hagermann A, Knollenberg J, Ball AJ, Banaszkiewicz M, Benkhoff J, Gadomski S, Gregorczuk W, Grygorczuk J, Hlond M, Kargl G, Kührt E, Kömle N, Krasowski J, Marczewski W, Zarnecki JC (2006) Muspus—a thermal and mechanical properties probe for the Rosetta Lander Philae. Space Sci Rev 128:339–362

Tsai L-W (1999) Robot analysis: the mechanics of serial and parallel manipulators. John Wiley & Sons, New York

Wood G, Kennedy D (2003) Simulating mechanical systems in simulink with SimMechanics. Technical report 91124v00, The MathWorks, Inc., Natick

Modeling and Simulation of Flapping Wings Entomopter in Martian Atmosphere

Adam Jaroszewicz, Jerzy Sąsiadek and Krzysztof Sibilski

Abstract This chapter presents modeling and simulation of a flapping wing Micromechanical Flying Insect (MFI) called Entomopter. This device is designed for sustained and autonomous flight in Martian atmosphere. The overall geometry of this micromechanical flying insect is based on hummingbirds and large insects. This chapter presents also a method for investigating the unsteady aerodynamics of flapping wings for micromechanical flying insect application. The experimental results show a good correlation with previously published data and the aerodynamic model compares favorably with the experimental results. The simulation results reveal important information regarding dynamics of the system, that could be used in future development work.

1 Introduction

The development of entomopter technology underwent rapid acceleration in last few years. Development of micro and nanotechnologies, coupled with computer aided design and manufacturing, as well as "more intelligent" software allows for significant number of new technologies and methods to be considered. Scientist began considering the possibility of building of miniature flying object *Micro Aerial Vehicle* (MAV), inspired by anatomy and aeromechanics of the birds and the insects flight. The name *"entomopter"* was created from two words coming from Greek language: *entomon*—insect, and *pteron*—wing. It is considered that entomopter is a kind of flying microrobots inspired with anatomy and mechanics of

A. Jaroszewicz (✉) · K. Sibilski
Wroclaw University of Technology, Wyb. Wyspianskiego 27 50-370 Wroclaw, Poland
e-mail: adam.jaroszewicz@pwr.wroc.pl

J. Sąsiadek
Carleton University, 1125 Colonel By Drive, Ottawa, ON K1S 5B6, Canada

J. Sąsiadek (ed.), *Aerospace Robotics*, GeoPlanet: Earth and Planetary Sciences,
DOI: 10.1007/978-3-642-34020-8_12,

the insects, and has dimensions no more than about 15 cm (6 inches) in three axes. It has high manoeuvrability, carries miniature devices and sensors, and is capable of transmitting required information (Jaroszewicz et al. 2007). From a biological point of view, flapping flight is one of the most fascinating forms of biological locomotion. Analogously to aquatic locomotion, it relies on the interaction of some undulatory motion of body appendages or the body itself with the surrounding fluid. However, unlike fish locomotion where the animal weight can be balanced by the body buoyancy, flying insects need to flap their wings continuously. In addition, insects can move in any direction in space, rotate their body rather easily, hover motionlessly, cruise at speeds of 1–10 ms, fly upside down or backward, and perform a 90° change of direction within 50 ms. This is remarkable when one considers that the wings appear to move almost symmetrically and with the same periodic pattern most of the times. Animalopter means a flying object constructed by man, which flies similarly to flying creatures (insects, birds and bats), i.e., by moving wings. It could be said that it is an *entomopter,* if it is an artificial insect, or an *ornitopter,* if it is an artificial bird. Wings of an animalopter are multifunctional devices, that create not only an aerodynamic lift, but also thrust, and, last but not least, can effectively control the flight (Dudley 2002; Maryniak and Sibilski 2008). Because of the complex equipment mounted on the animalopter it can be considered as a *flying microelectromechanical robot.* Animalopter has dimensions similar to the dimensions of a small bird (or a bat) or a large insect. Animalopter is more similar to a helicopter than to an aeroplane because of the complex motion of its body. Therefore, many concepts stemming from helicopter flight mechanics found its use in flight biomechanics, after taking into account animalopter‘s specificity (Jaroszewicz et al. 2007; Jaroszewicz and Sibilski 2010). The entomopter concept uses the same lift generating methods as insects do on Earth to generate lift within the Mars environment. Unlike aircraft or birds, insects generate lift by the continuous formation and shedding of vortices on their wings. This vortex formation and shedding produces very high wing lift coefficients, on the order of 5, compared to maximum lift coefficients of 1–1.2 for conventional airfoils. This very high lift generating capability is what allows insects to fly, hover and maneuver. It is believed that their ability to generate these high numbers for lift is related to the Reynolds numbers. As Reynolds number increases, the ability to fly is diminished. This high lift generating capability under low Reynolds number flight conditions poses an interesting solution for flight on Mars. Because of the low atmospheric density on Mars, a vehicle with a wing-span of 1 m would be in the same flight Reynolds number regime as most insects on Earth. Thus, it is conceivable to construct a vehicle that can fly near the surface of Mars (up to 100 m in altitude) while generating sufficient lift to allow it to fly slow, maneuver easily, and land safely. The dream of having the device with similar capabilities is the genesis behind the development of entomopter concept for Mars.

2 Aerodynamics of Flapping Wings

The aerodynamics of insect flight is substantially different from the aerodynamics associated to man-made fixed and rotary-winged vehicles. Traditional steady-state solutions of hydrodynamics equations fail to give reasonable answers. This inconsistency between theory and experimental evidence generated the famous paradox which states that according to laws of physics bumble bees cannot fly, but since they don't know that, they do fly. Reducing the size of the lifting surfaces and keeping the flight speed around 15 m/s (about 50 ft/s) makes the aerodynamic phenomena different from those found in normal size aircraft, mainly due to very low Reynolds number of the flow. Moreover, entomopter manoeuvring in this regime is subject to non-linear, unsteady aerodynamic loads (Dudley 2002; Jaroszewicz et al. 2007; Shyy 2008). The non-linearities and unsteadiness are due mainly to the large regions of 3-D separated flow and concentrated vortex flows that occur at large angles of attack. Accurate prediction of these non-linear, unsteady airloads is of great importance in the analysis of entomopter flight motion and in the design of its flight control system.

The unconventional aerodynamic concept associated with MAVs deserves a more detailed explanation. Insects fly by oscillating (frequency range: 5–200 Hz) and rotating their wings through large angles. This is possible because their wing articulation is not limited by an internal skeleton. The wing beat cycle can be divided into two distinct phases, the downstroke and the upstroke. At the beginning of downstroke the wing (as seen from the front of the insect) is in the uppermost position with the leading edge pointing forward. The wing is then pushed downwards and rotated continuously, resulting in large changes to the angle of attack. At the end of the downstroke the wing is twisted rapidly so that the leading edge points backwards, and the upstroke begins. During the upstroke the wing is pushed upwards and rotated again, changing the angle of attack throughout this phase. At the highest point the wing is twisted, so that the leading edge is pointing forwards again, and the next downstroke begins.

Figure 1 presents insect's wing tip trajectory. It isn't just a simple wing motion up and down, but it is much more complex. Such complex motion can be considerate as being composed of three different rotations: flapping, lagging and feathering. In forward flight the downstroke lasts longer than the upstroke, because of the need to generate thrust in addition to lift. In the hover, where lift only is required, the two strokes are of equal duration (Dudley 2002; Jaroszewicz et al. 2007; Sibilski 2004).

Detailed analyses of kinematics are central to an integrated understanding of animal flight. Four degrees of freedom in each wing are used to achieve flight in Nature: flapping, lagging, feathering, and spanning. Flapping is a rotation of the animal wing about the longitudinal axis of the animal body (this axis lies in the direction of the flight velocity), i.e., this is "up and down" motion. Lagging is a rotation about a "vertical" axis; this is, the "forward and backward" wing motion backward parallel to the body. Feathering is an angular movement about the wing

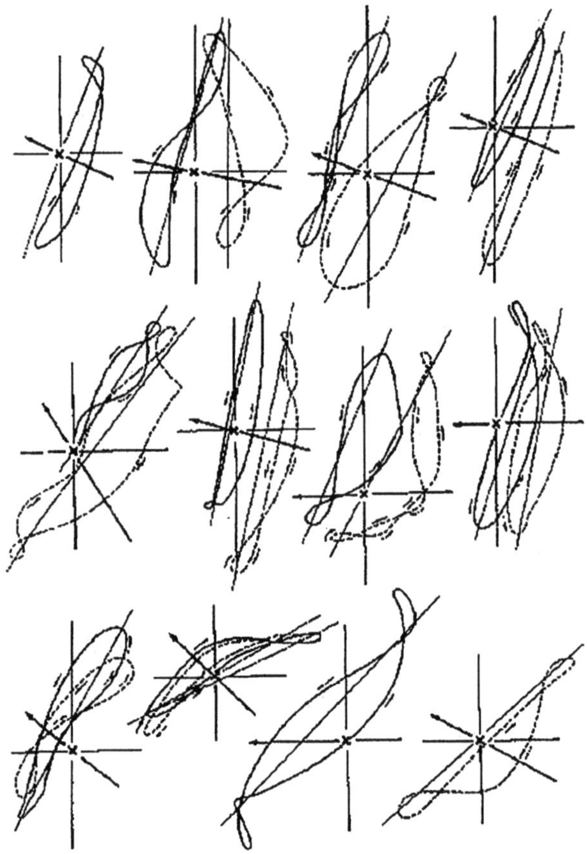

Fig. 1 Insect's wing tip trajectories (Jaroszewicz et al. 2007)

longitudinal axis (which may pass through the centre of gravity of the wing) which tilts the wing to change its angle of attack. Spanning is an expansion and contraction of the wingspan. Not all flying animals perform all of these motions. For instance, insects with low wing flap frequencies about 20 Hz (17–25 Hz) generally have very restricted lagging capabilities. Thus, flapping flight is possible with only two degrees of freedom: flapping and feathering. In the simplest physical models heaving and pitching represent these degrees of freedom. The motion of each bird wing may be decomposed into flapping, lagging, feathering (the rigid body motions) and also into more complex deflections of the surface from the base shape (vibration modes) (Jaroszewicz and Sibilski 2010; Maryniak and Sibilski 2008). Although numerical solutions of hydrodynamical equations are available today, a clear understanding of flapping flight aerodynamics has been obtained by dynamically scaled models of insect wings that can reproduce the same aerodynamics mechanisms as those in insect flight (Dudley 2002; Jaroszewicz et al. 2007). These experiments have unveiled three main aerodynamic mechanisms involved with flapping flight: delayed stall, rotational lift, and wake capture (Fig. 2).

- *Delayed Stall* (1): As the wing starts moving, a small vortex appears behind the leading edge, and an asymmetric, opposite swirl appears in the fluid close to the original resting position of the wing. The presence of two vortices moving in opposite directions but with identical strength is the equivalent principle of conservation of momentum for fluids. The vortex shedding process appears after the wing has traversed a distance of a few chord lengths; therefore, the increased aerodynamic force production can be captured only at the very beginning of the wing movement.
- *Rotational Lift* (2): Mechanism is the result of a combination of the translation and rotation of the wing. The magnitude of the aerodynamic force generated by the rotational lift is approximately proportional to the product of the angular velocity and translational velocity—Magnus Effect.
- *Wake Capture* (3): It is present at the beginning of each half-stroke after the wing has inverted its motion and started to move. The wake capture appears when the wing interacts with the effects of past strokes on the ambient fluid environment (Dudley 2002; Jaroszewicz et al. 2007).

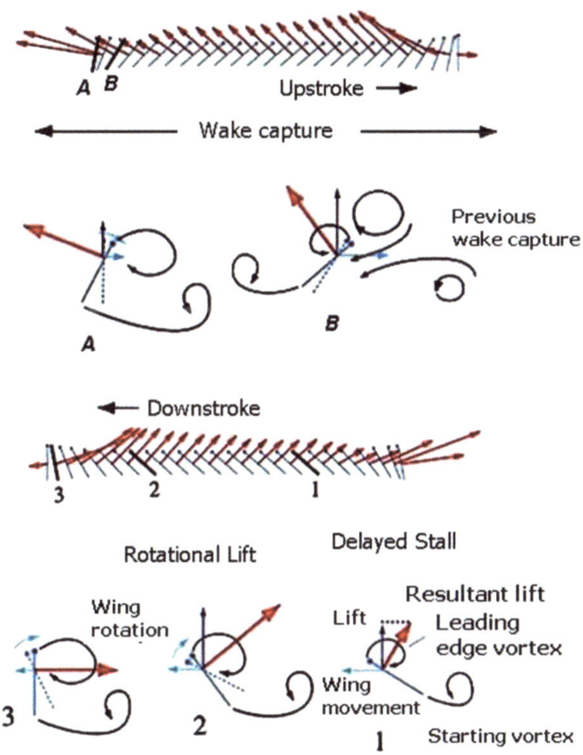

Fig. 2 Aerodynamics mechanism appearing in flapping wings (Jaroszewicz et al. 2007; Jaroszewicz and Sibilski 2010)

3 Definition of Aerodynamic Forces

3.1 Basic Assumptions

The model of the aerodynamics of entomopter (Fig. 3) presented in this chapter is a combination of the analytical model, created on the basis of equations for pseudo-steady flight with non-stationary aerodynamic effects, as well as the mathematical model for the flight buffeting based on results obtained experimentally from Robofly (Jaroszewicz et al. 2007; Khan 2005; Khan and Agrawal 2005).

Assumptions of the physical model of entomopter (Jaroszewicz et al. 2007; Jaroszewicz and Sibilski 2010):

- thorax and wings are not deformable;
- wings possess two degrees of the freedom and perform symmetrical movements buffeting and turn;
- aerodynamic forces generated by wings change periodically;
- thickness of the wing is a lot smaller than his spread and the string;
- changes of the angle of the turn of wings do not change positions of the mass centre and entomopter's moments of inertia.

Using the standard method *Blade Element Method* (BEM*)* the entomopter's wing is divided along the spread on elementary, infinitely thin, flat belts for which calculations are made of temporary elementary forces and aerodynamic moments. The model uses blade element analysis and takes into account unsteady effects such as leading edge vortex, rotational lift and virtual mass effects. Forces and moments working on the all wing area one receives across the integration of each elementary forces and moments along the spread of the wing (Fig. 4).

Based on a study of aerodynamics models presented in (Jaroszewicz and Sibilski 2010; Schenato et al. 2002), a total aerodynamics force is equal to a partial sum of steady and unsteady forces as follows:

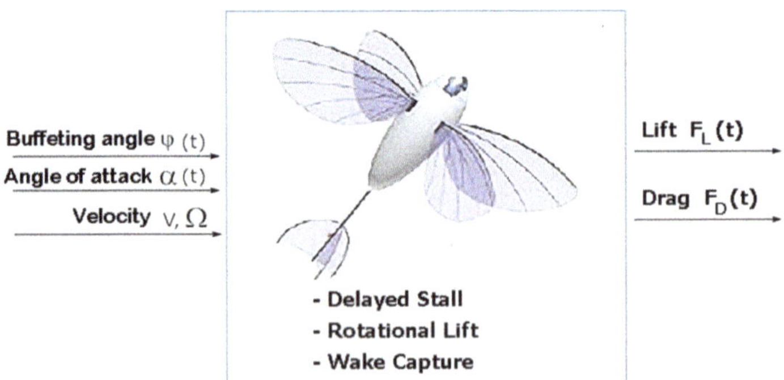

Fig. 3 Model of aerodynamics of entomopter (Jaroszewicz and Sibilski 2010)

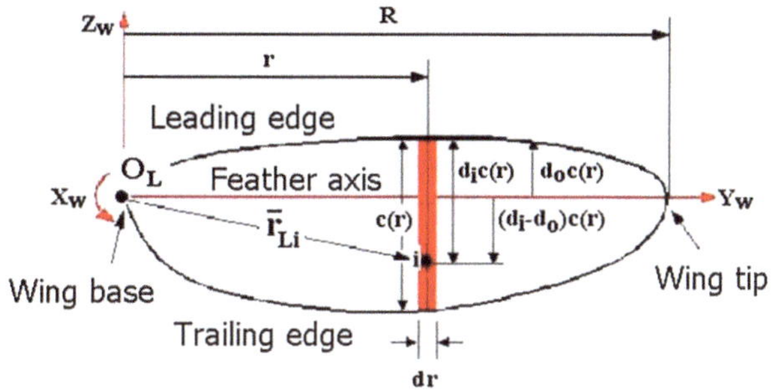

Fig. 4 The wing geometry of entomopter (Jaroszewicz and Sibilski 2010; Khan 2005)

$$F_{Total} = F_{QS} + F_{QUS} \tag{1}$$

where:

F_{QS} Quasi-Steady force.

F_{QUS} Quasi-Unsteady force.

The Quasi-Unsteady force is equal to a partial sum of forces: Leading Edge Vortex, Rotary, Virtual Mass as follows:

$$F_{QUS} = F_{LEV} + F_{Rot} + F_{VM} \tag{2}$$

where:

F_{LEV} Delayed Stall Force → shown as *Leading Edge Vortex*;

F_{Rot} Rotational Lift Force → shown as *Rotary*;

F_{VM} Wake Capture Force → shown as *Virtual Mass*.

3.2 Quasi–Steady Aerodynamic Forces

A quasi-steady state aerodynamic model assumes that force equations derived for 2D thin airfoils translating with a constant velocity $\bar{\mathbf{u}}_T$ and constant angle of attack α are as follows (Jaroszewicz et al. 2007; Schenato et al. 2002):

$$\begin{aligned} dF_N &= \frac{\rho}{2} C_N(\alpha) c(r) |\bar{\mathbf{u}}_T|^2 dr \\ dF_T &= \frac{\rho}{2} C_T(\alpha) c(r) |\bar{\mathbf{u}}_T|^2 dr \end{aligned} \tag{3}$$

where:

dF_N, dF_T Normal and tangential components of the aerodynamic force dF_A with respect to the airfoil profile.

α The angle of attack defined as the angle between the wing profile and the wing velocity relative to the fluid.

$c(r)$	Cord width of the aerofoil.
dr	Width of elementary, infinitely thin, flat belts of wing.
dr	Translational speed of wing.
$C_N(\alpha), C_T(\alpha)$	Dimensionless force coefficients.
ρ	Density of air.

An empirical approximation for the force coefficients is given by

$$C_N(\alpha) = 3,4\sin\alpha$$
$$C_T(\alpha) = \begin{cases} 0,4\cos^2(2\alpha) & 0 \leq \alpha \leq 45^o \\ 0 & \textit{for other angles} \end{cases} \tag{4}$$

These coefficients (4) have been obtained with RoboFly M.H Dickinson from Eq. (3) by measuring aerodynamic forces for different angles of attack and translational velocities (Jaroszewicz et al. 2007; Sane and Dickinson 2002).

The total aerodynamic force F_A decomposition in normal and tangential components is more intuitive, since aerodynamic forces are mainly pressure forces which act perpendicularly to the surface (Fig. 5). Lift and drag coefficient can be readily calculated from Eq. (5):

$$\begin{aligned} C_L(\alpha) &= C_N\cos\alpha - C_T\sin\alpha \\ C_D(\alpha) &= C_N\sin\alpha + C_T\cos\alpha \end{aligned} \tag{5}$$

Experimental research shows that the maximum lift coefficient is achieved for angles of attack of approximately 45°, considerably different from fixed and rotary wings which produce the maximum lift for angles of about 15° (Fig. 6).

In a steady state condition, the lift per unit length exerted on an aerofoil due to delayed stall is given by:

$$F_{tr,L} = \frac{\rho}{2}C_L(\alpha)\int_0^R c(r)|\bar{\mathbf{u}}_T|^2 dr \tag{6}$$

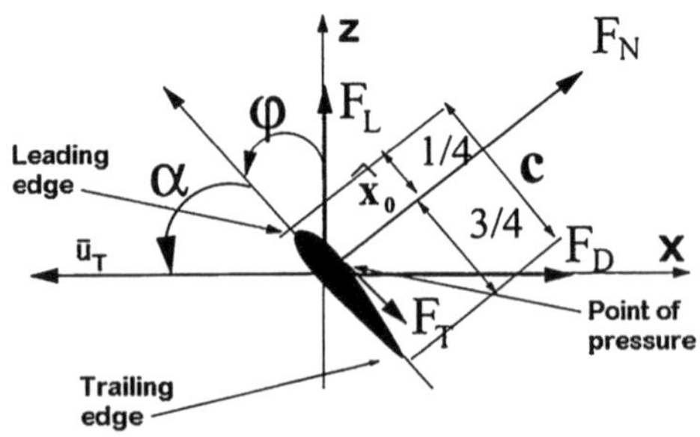

Fig. 5 Aerodynamic forces of the wing surface (Birch and Dickinson 2001; Jaroszewicz and Sibilski 2010)

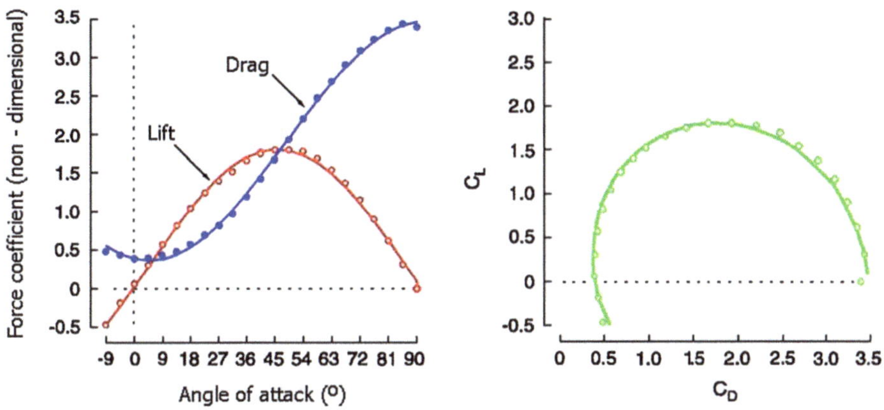

Fig. 6 Aerodynamic force coefficients and polar line empirically obtained from RoboFly data (Jaroszewicz and Sibiliski 2010; Schenato et al. 2002)

3.3 Leading Edge Vortex Force

Experimental research with models of insects' wings suggests that Leading Edge Vortex (LEV) force dF_{LEV} remains normal to the wing surface and it is not a function of wing rotation and acceleration. The LEV force value is proportional to the linear speed of the wing as follows:

$$dF_{LEV} = \frac{\rho}{2} C_1(\alpha) c(r) |\bar{\mathbf{u}}_T|^2 dr \tag{7}$$

where $C_1(\alpha)$ Dimensionless force coefficients, as a function of a local angle of attack.

The LEV force per unit length exerted on a airfoil due to delayed stall is given by (Jaroszewicz and Sibilski 2010; Khan 2005; Khan and Agrawal 2005):

$$F_{LEV} = \frac{\rho}{2} C_1(\alpha) \int_0^R c(r) |\bar{\mathbf{u}}_T|^2 dr = C_1 F_1 \tag{8}$$

3.4 Rotational Force

The aerodynamic force per unit length exerted on an aerofoil due to rotational lift is given by (Jaroszewicz and Sibilski 2010; Khan 2005; Khan and Agrawal 2005):

$$dF_{Rot} = \frac{\rho}{2} C_{Rot}(\alpha) c(r) |\bar{\mathbf{u}}_R|^2 dr \tag{9}$$

where:

$C_{Rot} = \pi\left(\frac{3}{4} - d_o c(r)\right)$	Rotational moment coefficients, approximately independent of the angle of attack.
$d_o c(r)$	Dimensionless distance of the rotation axis from the leading edge (for insect = 0.25).
$\bar{\mathbf{u}}_R$	Circuital velocity of the wing with respects to that axis y_{wL}.

The insect wing during the flight moves with the translational velocity $\bar{\mathbf{u}}_T$ and circuital velocity $\bar{\mathbf{u}}_R$. The total flow velocity $\bar{\mathbf{u}}$ can be written as a vector sum of the storage velocity. The rotational force could be presented as (Jaroszewicz and Sibilski 2010):

$$F_{Rot} = \frac{\rho}{2} C_{Rot}(\alpha) \int_0^R c(r) |\bar{\mathbf{u}}_R|^2 dr = C_2 F_2 \tag{10}$$

Rotational force coefficient C_2 is a dimensionless parameter whose theoretical value is dependent of location i and a vector of velocity $\bar{\mathbf{u}}$ along the chord of wing—$(d_i - d_0(r))$ as follows:

$$C_2(r) = d_i - d_0(r) \tag{11}$$

Aerodynamic quasi–steady forces generated on the wing surface are related between themselves. In the process of the modelling of aerodynamics of entomopter, it is assumed that the resultant aerodynamic forces, being a combination of two previously characterized aerodynamic forces, are generated at the wing. Leading Edge Vortex (LEV) Force and Rotational Force can be expressed as follows (see Fig. 7):

$$dF_{LEV+Rot} = \frac{\rho}{2} C_1(\alpha) c(r) |\bar{\mathbf{u}}|^2 dr \tag{12}$$

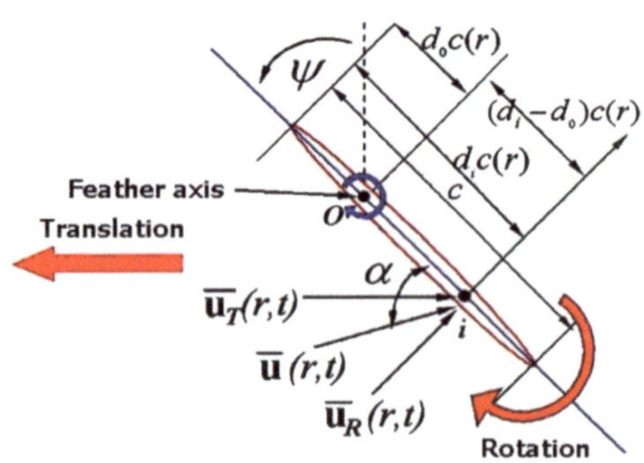

Fig. 7 Combined rotational and translational velocities—LEV effect (Jaroszewicz and Sibilski 2010; Khan 2005)

The resultant lift is given by (Sibilski 2004):

$$F_{LEV+Rot} = \frac{\rho}{2} C_1 \int_0^R c(r) |\bar{\mathbf{u}}|^2 dr \tag{13}$$

3.5 Wake Capture Force

Simulation of the Wake Capture effect is difficult because an analytical model is complex. However, the so called Virtual Mass possibly becomes an estimation of the value for generated aerodynamic force. As the wing accelerates, it moves along some mass of air, assumed to be contained in a cylinder with diameter equal to the chord. The rotational trace was simulated as the whirl of air in the shape of the cylinder with diameter equal to the length of the string of the wing (Fig. 8).

Wake Capture Force (*Virtual Mass Force)* is acting on a section located at a distance r from the flapping axis. The acceleration of this mass of air shows up as a virtual mass force and is given by (Jaroszewicz and Sibilski 2010; Khan 2005):

$$dF_{VM} = dm\dot{\mathbf{u}}_n(r,t) \tag{14}$$

where:

$\dot{\mathbf{u}}_n(r,t)$ Rate of change of normal velocity component at the mid-chord location in the wing frame.

dm Mass of air enclosed in a thin cylinder of width ***dr*** and a diameter equal to the chord c(r) at a distance r from the flapping axis.

The mass of air is $\frac{\rho\pi c^2}{4}$, and therefore Eq. (14) can be written as (Jaroszewicz and Sibilski 2010; Khan 2005):

$$dF_{VM} = C_3 \frac{\rho\pi}{4} \dot{\mathbf{u}}_n(r) c(r)^2 dr \tag{15}$$

The resultant total virtual mass force is given by (Jaroszewicz and Sibilski 2010; Khan 2005):

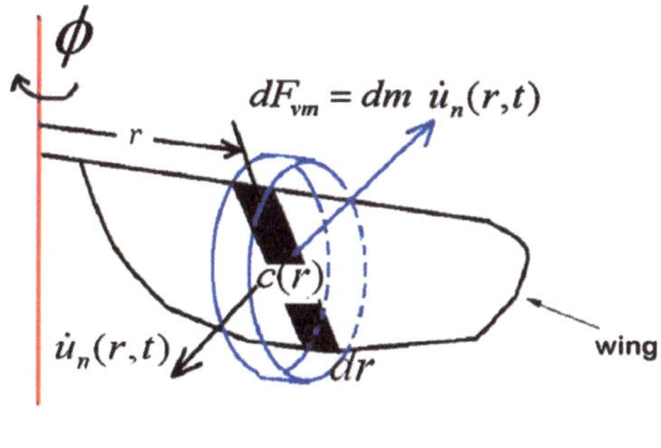

Fig. 8 Virtual mass force (Jaroszewicz and Sibilski 2010; Khan and Agrawal 2005)

$$dF_{VM} = C_3 \frac{\rho\pi}{4} \dot{\mathbf{u}}_n(r) c(r)^2 dr \tag{16}$$

where:
C_3 Coefficient of virtual mass effect.

3.6 Total Aerodynamic Forces

The total aerodynamic forces used for simulations throughout this chapter are based on Eq. (5), and therefore the normal F_N and tangential F_T forces are (Jaroszewicz and Sibilski 2010; Khan 2005):

$$\begin{aligned} F_N &= C_N F_1 + C_1 F_1 + F_2(C_1, C_2) + C_3 F_3 \\ F_T &= C_T F_1 \end{aligned} \tag{17}$$

The total lift F_L and drag F_D forces acting on the wing can be derived through a trigonometric transformation analogous to the one used in Eq. (5) as follows:

$$F_L = F_N \cos\alpha(t) - F_T \sin\alpha(t) \tag{18}$$

$$F_D = F_N \sin\alpha(t) + F_T \cos\alpha(t) \tag{19}$$

3.7 Equations of Motion

The formalism of analytical mechanics allows to present dynamic equations of motion of mechanical systems in generalized co-ordinates, giving a useful tool for development of equations of motion for aircraft. Good examples are Gibbs-Appel equations. Those equations have the following form (Goraj and Pietrucha 1998):

$$\frac{d}{dt}\left(\frac{\partial S}{\partial \ddot{\mathbf{q}}}\right) = \mathbf{Q} \tag{20}$$

where $\mathbf{q}$ is the vector of generalized co-ordinates; $S(\mathbf{q}, \dot{\mathbf{q}}, \ddot{\mathbf{q}}, t)$ is the so-called Appel function, or functional of accelerations. Functional S for i-th element of the mechanical system is given by the equation (Sibilski 1998)

$$S = \frac{1}{2} \iiint_V \dot{\mathbf{v}}^i \circ \dot{\mathbf{v}}^i dm^i \tag{21}$$

where $\dot{\mathbf{v}}^i$ means the vector of absolute acceleration of elementary mass dm^i of i-th body of the dynamical system considered (Fig. 1).

$$\dot{\mathbf{v}}^i = \dot{\mathbf{v}}_0^i + \varepsilon_0^i \times \boldsymbol{\rho}^i + \boldsymbol{\omega}_0^i \times \left(\boldsymbol{\omega}_0^i \times \boldsymbol{\rho}^i\right) \tag{22}$$

where ρ^i is the radius vector defining position of i-th element of entomopter body in inertial system of co-ordinates (Fig 9):

Defining the inertia matrix:

$$\mathbf{M}^i = \begin{bmatrix} m^i\mathbf{I} & m^i\tilde{\rho}^{iT} \\ m\tilde{\rho}^i & \mathbf{J}_O^i \end{bmatrix} \tag{23}$$

and introducing the vector defining the dynamical forces (effects of centrifugal, gyroscopic, and Coriolis forces):

$$\mathbf{h}^i = \begin{bmatrix} m^i\tilde{\omega}_0\mathbf{v}_0^i + m^i(\tilde{\omega}_0)^2\rho^i \\ m^i\tilde{\rho}^i\tilde{\omega}_0\mathbf{v}_0^i + \tilde{\omega}_0\mathbf{J}_0^i\boldsymbol{\omega}_0^i \end{bmatrix} \tag{24}$$

where m^i is the mass of the i-th element, $\mathbf{J}_O^i$ the tensor of inertia of the i-th element, ω_0 the vector of the angular velocity, $\mathbf{v}_0^i$ the vector of the velocity of the i-th element, and assuming that:if: $\mathbf{a} = \left[a_\xi, a_\eta, a_\varsigma\right]^T$, then $\tilde{\mathbf{a}} = \begin{bmatrix} 0 & -a_\varsigma & a_\eta \\ a_\eta & 0 & -a_\xi \\ -a_\eta & a_\xi & 0 \end{bmatrix}$;

the term (2) can be expressed in the following matrix form:

$$S = \frac{1}{2}\left[\dot{\mathbf{v}}^i + \left(\mathbf{M}^i\right)^{-1}\mathbf{h}^i\right]^T \mathbf{M}^i\left[\dot{\mathbf{v}}^i + \left(\mathbf{M}^i\right)^{-1}\mathbf{h}^i\right] \tag{25}$$

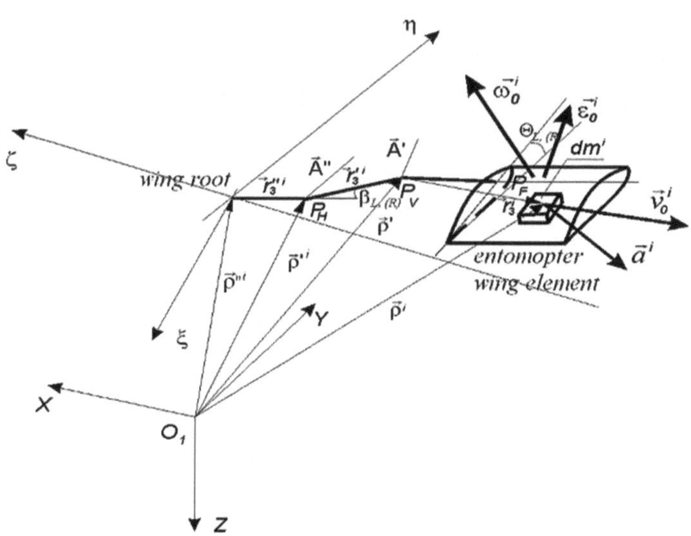

Fig. 9 Location of points, radius vectors and vectors of velocities and accelerations on Entomopter wing element

Calculating the matrixes $\mathbf{M}^i$, $\mathbf{h}^i$, the Appel function S^i for all k bodies of the system, and defining the matrixes:

$$\mathbf{M} = diag\left[\mathbf{M}^1, \mathbf{M}^2, \ldots \ldots, \mathbf{M}^k\right] \tag{26}$$

$$\mathbf{v} = \left[\left(\mathbf{v}^1\right)^T, \left(\mathbf{v}^2\right)^T, \ldots \ldots, \left(\mathbf{v}^k\right)^T\right]^T \tag{27}$$

and

$$\mathbf{h} = \left[\left(\mathbf{h}^1\right)^T, \left(\mathbf{h}^2\right)^T, \ldots \ldots, \left(\mathbf{h}^k\right)^T\right]^T \tag{28}$$

functional S for the whole mechanical system is given by the equation:

$$S = \frac{1}{2}\left(\dot{\mathbf{v}} + \mathbf{M}^{-1}\mathbf{h}\right)^T \mathbf{M}\left(\dot{\mathbf{v}} + \mathbf{M}^{-1}\mathbf{h}\right) \tag{29}$$

Assuming that $\mathbf{q}$ is a vector of generalized coordinates for mechanical system, the relations between $\mathbf{q}$ and $\mathbf{v}$ are given by the equation:

$$\mathbf{v} = \mathbf{D}(\mathbf{q}, t)\dot{\mathbf{q}} + \mathbf{f}(\mathbf{q}, t) \tag{30}$$

hence:

$$\dot{\mathbf{v}} = \mathbf{D}(\mathbf{q}, t)\ddot{\mathbf{q}} + \varphi(\mathbf{q}, \dot{\mathbf{q}}, t) \tag{31}$$

where:

$$\varphi = \dot{\mathbf{D}}\dot{q} + \dot{\mathbf{f}}$$

Therefore, the Appel function can be expressed by the following relation:

$$S(\mathbf{q}, \dot{\mathbf{q}}, \ddot{\mathbf{q}}, t) = \frac{1}{2}\left(\mathbf{D}\ddot{\mathbf{q}} + \varphi + \mathrm{M}^{-1}\mathbf{h}\right)^T \mathbf{M}\left(\mathbf{D}\ddot{\mathbf{q}} + \varphi + \mathrm{M}^{-1}\mathbf{h}\right) \tag{32}$$

Assuming, that:

$$\mathbf{M}_g = \mathbf{D}^T\mathbf{M}\mathbf{D} \text{ and } \mathbf{h}_g = \mathbf{D}^T(\mathbf{M}\varphi + \mathbf{h})$$

and considering that:

$$\mathbf{D}\left(\mathbf{D}^T\mathbf{M}\mathbf{D}\right)^{-1}\mathbf{D}^T = \mathbf{M}^{-1}$$

Eq. (14) can be expressed in the form:

$$S(\mathbf{q}, \dot{\mathbf{q}}, \ddot{\mathbf{q}}, t) = \frac{1}{2}\left(\ddot{\mathbf{q}} + \mathrm{M}_g^{-1}\mathbf{h}_g\right)^T \mathbf{M}\left(\ddot{\mathbf{q}} + \mathrm{M}_g^{-1}\mathbf{h}_g\right) \tag{33}$$

In case of a flapping wings MAV treated as a mechanical system containing rigid fuselage and two rigid wings fixed to the fuselage with two hinges, the vector generalized co-ordinates have the following form:

$$\mathbf{q} = [x_s, y_s, z_s, \Phi, \Theta, \Psi, \ \beta_L, \beta_R, \theta_L, \theta_R]^T \tag{34}$$

the vector of quasi-velocities can be expressed by the following equation

$$\mathbf{w} = \left[U, V, W, P, Q, R, \ \dot{\beta}_L, \dot{\beta}_R, \dot{\theta}_L, \dot{\theta}_R\right]^T \tag{35}$$

For the holonomic dynamical system the relation between generalized velocities $\dot{\mathbf{q}} = [\dot{q}_1, \dot{q}_2, \ldots\ldots\ldots \dot{q}_n]$ and quasi velocities $\mathbf{w} = [w_1, w_2, \ldots\ldots\ldots w_n]$ is as follows:

$$\dot{\mathbf{q}} = \mathbf{A}_T(\mathbf{q})\mathbf{w} \tag{36}$$

The matrix $\mathbf{A}_T$ has a construction:

$$\mathbf{A}_T = \begin{bmatrix} \mathbf{A}_G & \mathbf{0} & \mathbf{0} \\ \mathbf{0} & \mathbf{C}_T & \mathbf{0} \\ \mathbf{0} & \mathbf{0} & \mathbf{I} \end{bmatrix} \tag{37}$$

The matrices $\mathbf{A}_G$ and $\mathbf{C}_T$ are classical matrices of transformations for kinematics relations (Willmott and Ellington 1997), the unit matrix $\mathbf{I}$ has dimension 4×4.

From (36) the following relation can be written:

$$\ddot{\mathbf{q}} = \mathbf{A}_T(\mathbf{q})\dot{\mathbf{w}} + \dot{\mathbf{A}}_T\mathbf{w} \tag{38}$$

Finally, the Appel function has the following form:

$$S^*(\mathbf{q}, \mathbf{w}, \dot{\mathbf{w}}, t) = \frac{1}{2}\left(\dot{\mathbf{w}} + \mathbf{M}_w^{-1}\mathbf{h}_w\right)^T \mathbf{M}_w\left(\dot{\mathbf{w}} + \mathbf{M}_w^{-1}\mathbf{h}_w\right)$$

where $\mathbf{M}_w(\mathbf{q}) = \mathbf{A}_T^T\mathbf{M}_q\mathbf{A}_T$, and: $\mathbf{h}_w(\mathbf{q}, \mathbf{w}) = \mathbf{A}_T^T\left(\mathbf{M}_q\dot{\mathbf{A}}_T + \mathbf{h}_q\right)$

At last the Gibbs-Appel equations of motion, written in quasi velocities have the following form:

$$\left(\frac{\partial S^*}{\partial \dot{\mathbf{w}}}\right)^T = \left[\frac{\partial S^*}{\partial \dot{w}_1}, \ldots\ldots\ldots\ldots, \frac{\partial S^*}{\partial \dot{w}_k}\right]^T = \mathbf{M}_w(\mathbf{q}, t)\dot{\mathbf{w}} + \mathbf{h}_w(\mathbf{q}, \mathbf{w}, t) = \mathbf{Q}^*(\mathbf{q}, \mathbf{w}, t) \tag{39}$$

The vector $\mathbf{Q}^*$ is the sum of aerodynamic loads, potential forces acting on the MAV, and other non-potential forces.After some elementary conversions, the equations of motion can be presented in the following matrix form:

$$\dot{\mathbf{V}} = \mathbf{M}^{-1}\left\{\mathbf{F} + \mathbf{G} - \left[\mathbf{M}\left(\dot{\mathbf{J}}_\Omega + \mathbf{J}_\Omega\mathbf{J}_\Omega\right)\mathbf{R}_C + \mathbf{M}\mathbf{J}_\Omega\dot{\mathbf{R}}_C + \mathbf{M}\mathbf{J}_\Omega\right]\right\} \tag{40}$$

$$\dot{\Omega} = \mathbf{J}_B^{-1}\left\{\mathbf{M}_0 + \mathbf{R}_C \times \mathbf{G} - \begin{bmatrix} \mathbf{J}_B^R\dot{\mathbf{O}}_R + \mathbf{J}_B^L\dot{\mathbf{O}}_L + \left(\mathbf{J}_s + \mathbf{J}_S^w + 2\mathbf{J}_\Omega\mathbf{J}_S + \dot{\mathbf{J}}_S^w\right)\mathbf{V} \\ +\left(\dot{\mathbf{J}}_B^R + \mathbf{J}_\Omega\mathbf{J}_B^R\right)\mathbf{O}_R + \left(\dot{\mathbf{J}}_B^L + \mathbf{J}_\Omega\mathbf{J}_B^L\right)\mathbf{O}_L + \mathbf{J}_\Omega\mathbf{J}_B\Omega \end{bmatrix}\right\} \tag{41}$$

where: $\mathbf{M} = m\mathbf{I}$, m—mass of entomopter, $\mathbf{I}$—unit matrix, $\mathbf{F} = [F_x, F_y F_z]^T$—vector of aerodynamic forces, $\mathbf{M}_0 = [L,M,N]^T$—vector of aerodynamic moments, $\mathbf{V} = [U,V,W]^T$—velocity vector; $\mathbf{\Omega} = [P,Q,R]^T$—vector of angular velocity, $\mathbf{O}_R = \left[P - \dot{\beta}, Q + \dot{\theta}, R\right]^T$ —vector of right wing angular rates, $\mathbf{O}_L = \left[P + \dot{\beta}, Q + \dot{\theta}, R\right]^T$ —vector of left wing angular rate; $\mathbf{R}_c = [x_c, y_c, z_c]^T$ vector of the center of mass; $\mathbf{J}_\Omega = \begin{bmatrix} 0 & -R & Q \\ R & 0 & -P \\ -Q & P & 0 \end{bmatrix}$, $\mathbf{J}_S = \begin{bmatrix} 0 & -S_z & S_y \\ S_z & 0 & -S_x \\ -S_y & S_x & 0 \end{bmatrix}$; where: S_x, S_y, S_z – static moments entomopter without wings; $\mathbf{J}_S^w$ —matrix of static moments of MAV's wing; $\mathbf{J}_B$—inertial moment of entomopter without wings; $\mathbf{J}_B^R$, $\mathbf{J}_B^L$ —inertial moments of right and left wing, respectively. The control vector is defined as follows:

$$\mathbf{u} = [\beta\,\theta\,\omega\,\psi]^T \tag{42}$$

where: β—flapping angle of wings; θ—feathering angle of wings, ω—frequency of wing motion in respect to the body; ψ—phase shifting between feathering and flapping; and: $\beta = \beta_0 \sin\omega t$, $\dot{\beta} = \beta_0\omega \sin\omega t$, $\theta = \theta_0 \sin(\omega t + \psi)$ and $\dot{\theta} = \theta_0\omega\cos(\omega t + \psi)$.

3.8 Synthesis of Nonlinear Control Law

Synthesis of nonlinear control law was performed on the basis of nonlinear inverse dynamics. It requires the following form of mathematical model of motion:

$$\dot{\mathbf{w}} = \mathrm{f}(\mathbf{w}) + \mathrm{g}(\mathbf{w})\mathbf{u}, \quad \mathbf{y} = \mathbf{h}\,(\mathbf{w}) \tag{43}$$

where: the dimension of vector of state $\mathbf{x}$ equals n, and the dimension of vector of control $\mathbf{u}$ and vector of output $\mathbf{y}$ equals m. Control law is in the following form:

$$\mathbf{u} = \mathbf{D}^{-1}(\mathbf{w})[\mathbf{v} - \mathbf{N}(\mathbf{w})] \tag{44}$$

where: $\mathbf{v} = \mathbf{P}_0\mathbf{y}_z - \sum_{j=0}^{r-1}\mathbf{P}_j\mathbf{y}^{(j)}$. Index j means j-th derivative of the vector of output.

$\mathbf{P}_j$ are constant matrices with dimensions $m \times m$, assumed arbitrarily, and $\mathbf{y}_z$ means the given motion. Scattering matrix $\mathbf{D(x)}$ is in the following form:

$$\mathbf{D} = \begin{bmatrix} L_{\mathbf{G}1} L_{\mathbf{T}}^{r_1 - 1} h_1, & \ldots, & L_{\mathbf{G}1} L_{\mathbf{T}}^{r_1 - 1} h_1 \\ \ldots & \ldots & \ldots \\ L_{\mathbf{G}1} L_{\mathbf{T}}^{r_m - 1} h_m, & \ldots, & L_{\mathbf{G}m} L_{\mathbf{T}}^{r_m - 1} h_m \end{bmatrix} \tag{45}$$

in which $L_{\mathbf{F}}\mathrm{h} = \nabla \mathrm{h}\mathbf{F}$ is a Lie derivative of function h in relation to vector field $\mathbf{F}$ and, $\mathbf{N}(\mathbf{w}) = \left[L_{\mathbf{F}}^{r_j} h_j(\mathbf{w})\right]$, dim $\mathbf{w} = n$, dim $\mathbf{u} =$ dim $\mathbf{y} = m$, and the degree of relativity $r = \sum_{j=1}^{m} r_j$ exists if det$\mathbf{D}(\mathbf{w}) \neq 0$; and

$$L_{Gi} L_F^k \mathbf{h}_j(\mathbf{w}) = 0\,, \quad \text{for} \quad 0 \le k \le r_i - 1 1 \le i, j\, 0 \le m.$$

Control process will be realised if the degree of relativity $r = n$. In the case when $r \le n$ success of control depends on internal dynamics.

4 Simulation Flight Dynamics of Entomopter

Figures 10 and 11 show the simulated aerodynamic forces for a typical wing motion of insect wing (*entomopter*).

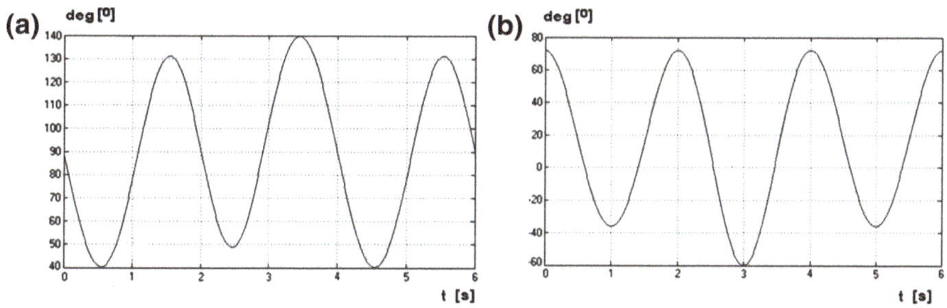

Fig. 10 Visualization of changes of the angle of attack in time (**a**) angle of the buffeting (**b**) angle of the entomopter wing

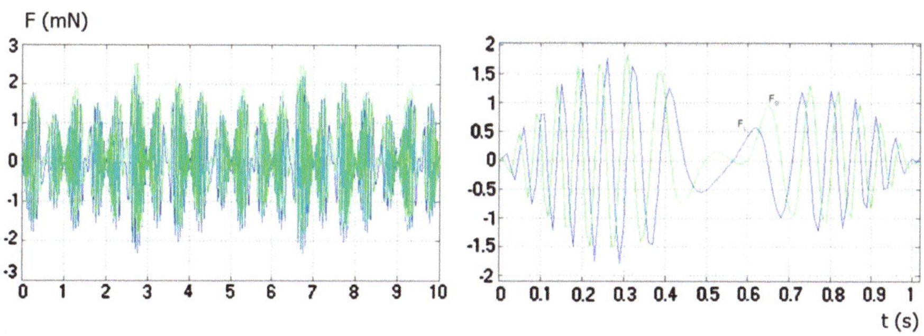

Fig. 11 Visualization of changes of the lift F_L and drag F_D forces in time (Eqs. 18 and 19)

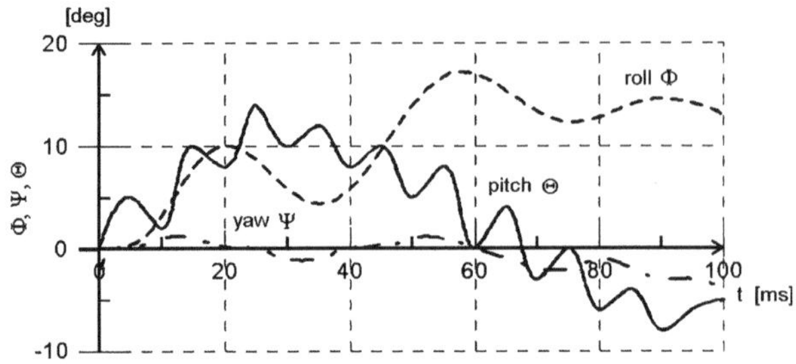

Fig. 12 Hovering stabilization—angular position of entomopter

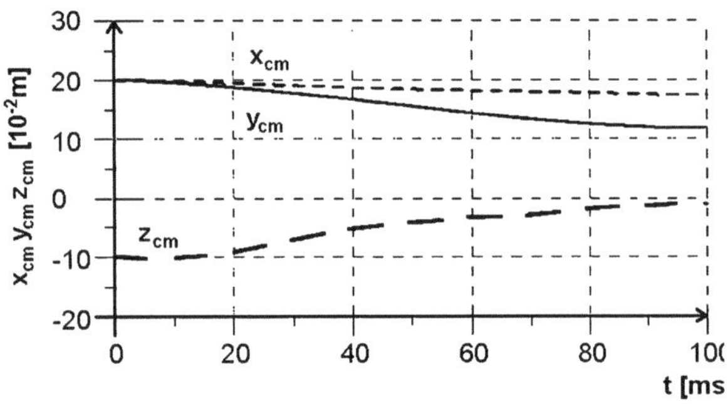

Fig. 13 Hovering stabilization—position of centre of mass (x_{cm} y_{cm} z_{cm}) of entomopter

In Figs. 12, 13 and 14 some results of simulations of hovering stabilization are presented.

Figures 12 and 13 show the resulting position and velocity trajectories, together with the corresponding parameters chosen at each wingbeat of hovering stabilization. The linear displacements are recovered from 20 to 100 cm to its equilibrium point within 100 ms.

Figure 14 shows the proposed control vectors in simulated wings stroke of entomopter within 50 wing strokes.

5 Conclusions

This chapter presents a methodology for the theoretical determination of steady and unsteady aerodynamic force coefficients based on the principle of dynamic similarity. Simulations performed indicate that changes of the lift and drag forces for the buffeting movement of entomopter wings had an oscillatory character. Despite some small discrepancies due to the inaccuracies in modeling of the wake capture forces, the mathematical model presented in this chapter accurately predicts motions of from flying insect. Based on the results obtained in this study, it could be concluded that the proposed model can be used in the aerodynamic module for the determination of aerodynamic force and moment components of entomopter.

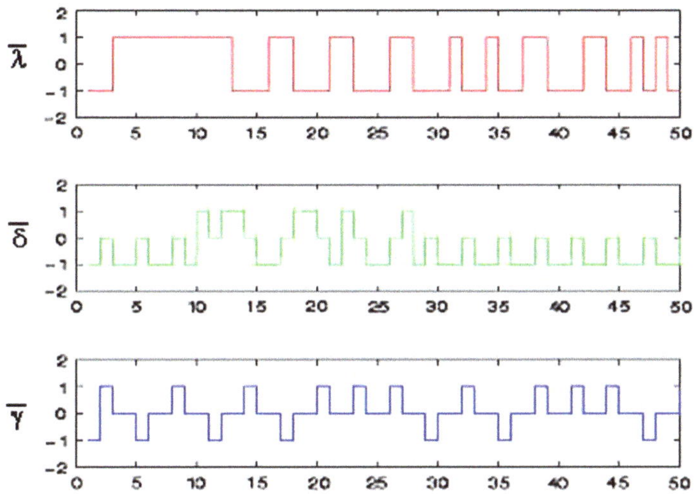

Fig. 14 Stabilization of hovering flight—control vectors (values of angles in relation to amplitudes of oscillations): $\bar{\lambda} = \lambda/\lambda_0, \quad \bar{\delta} = \delta/\delta_0, \quad \bar{\gamma} = \gamma/\gamma_0$

Probably the most difficult entomopter system is the flight control system, which should by highly autonomous and should operate instantaneously. It should be noted that there are relatively strong forces and moments caused by laminar flow. Moreover, it is very difficult to foresee the conditions in which the flight will take place. Small mass and dimensions, and therefore (moments of inertia), the unsteady effects of flow caused by gushes of the air and manoeuvres will significantly influence the aerodynamic loads of the entomopter. This is obvious because of extremely low unit load of lifting surface of this aircraft.

References

Birch JM, Dickinson MH (2001) Spanwise flow and the attachment of the leading-edge vortex on insect wings. Nature 412(6848):729–733

Dudley R (2002) The biomechanics of insect flight. Princeton University Press, New Jersey

Goraj Z, Pietrucha J (1998) Basic mathematical relations of fluid dynamics for modified panel methods. J Theor App Mech 36(1):47–66

Jaroszewicz A, Sibilski K (2010) Modeling and simulation of entomopter flight dynamics, 4th international conference on SAUAV—2010, Suchedniów

Jaroszewicz A, Sibilski K, Sibilska A, Żyluk A (2007) Biomimic sensors guided flight stability and control for flapping wings autonomous micro air vehicle (Entomopter), AIAA 2007, p 664

Khan ZA(2005) Modeling and simulation of flapping wing micro air vehicles, IDETC/CIE—ASME

Khan ZA, Agrawal SK (2005) Wing force and moment characterization of flapping wings for micro air vehicle application. IEEE 3:1515–1520
Maryniak J, Sibilski K (2008) Microelectromechanical flying vehicles. Mech Aviation ML-XIII 1:219–252
Sane SP, Dickinson MH (2002) The aerodynamic effects of wing rotation and a revised quasisteady model of flapping flight. J Exp Biol 205:10871096
Schenato L, Wu WC, Sastry SS (2002) Attitude control for a micromechanical flying insect via sensor output feedback, IEEE Trans Robot Autom, 7th Intern. Conf. Control Automation Robotics and Vision. ICARCV 2002, Vol 2. Singapore, pp 1031–1036
Shyy W (2008) Aerodynamics of low reynolds number flyers, Cambridge University Press, Cambridge
Sibilski K (1998) Modeling of an agile aircraft flight dynamics in limiting flight conditions. Military University of Technology, Warsaw
Sibilski K (2004) Influence of insect morphology on MAV's design scheme, international conference on SAUAV—2004, Kielce
Willmott AP, Ellington CP (1997) The mechanics of flight in the Hawkmoth Manduca Sexta. Part I. Kinematics of hovering and forward flight. J Exp Biol 200:2705–2722

The Experimental Results of the Functional Tests of the Mole Penetrator KRET in Different Regolith Analogues

Karol Seweryn, Marek Banaszkiewicz, Stanisław Bednarz, Monika Ciesielska, Andrzej Gonet, Jerzy Grygorczuk, Tomasz Kuciński, Tomasz Rybus, Mirosław Rzyczniak, Roman Wawrzaszek, Łukasz Wisniewski and Maciej Wójcikowski

Abstract Depending on the specific region, the unmanned exploration of the planetary bodies can be divided into three groups: operations above the surface, operations on the surface and operations under the surface. In this chapter we will focus on the operations under the surface, connected with them requirements, available technology and possible output from such research. In this context we will present mole KRET device as a one of the possible solutions for low power consuming device which can be treated as a sub-surface end-effector of a more complicated robotic system. The experimental results of the functional tests of the mole in a 5 m test-bed system will be provided for different regolith analogue. The detailed investigation of lunar analogue will show how the progress of the mole depends on the compaction ratio of the material.

1 Introduction

The mole penetrator KRET belongs to the family of the Low Velocity Penetrator (LVP) driven by hammering actions. It is a low velocity, medium to high stroke energy but low power, self-driven penetrator, designed as a carrier of different sensors for in situ investigations of subsurface layers of planetary bodies.

Maciej Wójcikowski—Deceased

K. Seweryn (✉) · M. Banaszkiewicz · M. Ciesielska · J. Grygorczuk · T. Kuciński · T. Rybus · R. Wawrzaszek · Ł. Wisniewski
Space Research Centre of the Polish Academy of Sciences, 18a Bartycka Street 00-716 Warsaw, Poland
e-mail: kseweryn@cbk.waw.pl

S. Bednarz · A. Gonet · M. Rzyczniak · M. Wójcikowski
AGH University of Science and Technology, Mickiewicza 30 Street 30-059 Krakow, Poland

J. Sąsiadek (ed.), *Aerospace Robotics*, GeoPlanet: Earth and Planetary Sciences,
DOI: 10.1007/978-3-642-34020-8_13,

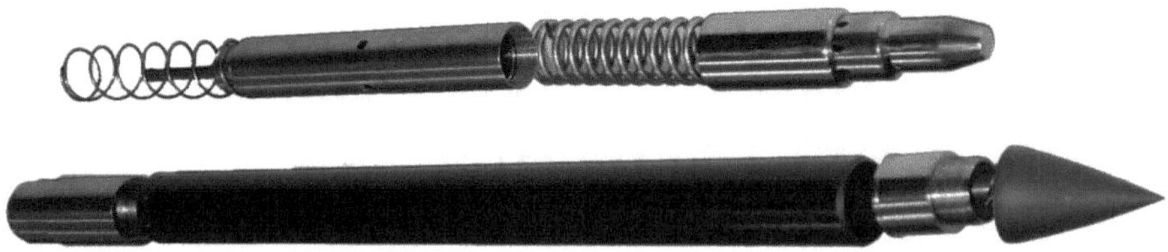

Fig. 1 Mole penetrator KRET. The internal and external parts are on the *top* and *bottom*, respectively

The basic principle of operation of such a penetrator was presented in Grygorczuk et al. (2011a), whereas detailed comparison between results and theoretical model was shown in technical reports and summarized in this paper. The maximum insertion depth is limited by energy of mole's single stroke and soil resistance for the dynamic penetration.

A mole penetrator KRET (Fig. 1) has been designed, developed, and successfully tested at Space Research Centre PAS in Poland (Grygorczuk et al. 2008). The KRET design takes advantage of the multi-purpose sensors for surface and subsurface science (MUPUS) penetrator (Grygorczuk et al. 2007)—a payload of Philae lander on Rosetta mission (Spohn et al. 2007).

The parameters of the mole KRET are listed in the Table 1.

There are several functions of the mole penetrator:

- Indirect measurements of the mechanical parameters of the regolith by exploitation of the progress of the mole in the regolith as well as using seismic while penetrating (SWP) technique which is equivalent of the seismic while drilling technique (Coste et al. 2010) for penetrators.
- Transporting sensors below the surface: for example, a heat flux sensor (Banaszkiewicz et al. 2007), a camera or even a small spectrometer (Tulej et al. 2011).
- Sampling the material and transporting it to the surface (Grygorczuk et al. 2011b).
- If considering a group of penetrometers, it can allow for a series of more sophisticated research, e.g., the comparison of regolith properties at different depths, etc. (Dąbrowski and Banaszkiewicz 2006).

The work presented in this chapter provides the results from the tests of the mole penetrator KRET in dry quartz sand and newly developed lunar regolith analogue AGK-2010 (created in cooperation between SRC PAS and AGH University of

Table 1 Major parameters of the mole KRET devices

Outer diameter (mm)	20.4
Length (mm)	336
Total mass (g)	500
Energy of the driving spring (J)	2.2
Average/Peak power consumption (W)	0.28/0.7

Science and Technology, patent application no. P397651) in new test-bed system allowing for tests to be performed at the depth of up to 5 m (Grygorczuk et al. 2011a). The dynamics of the mole during one stroke are also presented.

2 Operation Principle

The operational principle of the mole KRET was presented in several of our previous chapters (Grygorczuk et al. 2009) but for better understanding the presented results it is summarized in this chapter.

The principle of operation of the mole bases on the interaction between three masses: the inserted cylindrical casing, the hammer, and the rest of the mass, acting as a support mass. Additionally, the driven spring should act on the hammer and the support, and the return spring should act on the support and the casing.

In a single work cycle (one stroke) of the mole KRET, four phases can be distinguished (Fig. 2):

- **PHASE 1**: Driven hammer compresses the driven spring.
- **PHASE 2**: Released hammer accelerates and hits the casing. As a result of exchange of energy and momentum, the casing is inserted at Δx_1. The support moves in the opposite direction.
- **PHASE 3**: The support reaches the highest position compressing the return spring.
- **PHASE 4**: The support accelerated by the return spring and gravity hits the casing and causes its additional move. Total progress of insertion is Δx_2.

3 The Measurement System Principle

The progress of the mole penetrator in granular medium depends on the mechanical properties of this medium. To ensure conditions comparable to the ones on the Moon, a new analogue of the Moon's regolith was developed. This

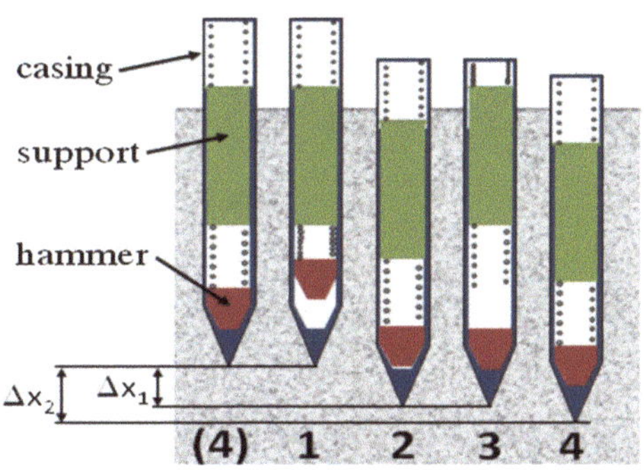

Fig. 2 Schematic principle of operation of the mole

analogue, called AGK-2010 (Seweryn et al. 2011), is based on the particle size distribution of CHENOBI lunar highlands physical regolith simulant and has the same mechanical properties (e.g., shear durability). The comparison of the new lunar analogue with the existing one is presented in Fig. 3.

The 5 m test-bed consists of a free-standing anchored 6.5 m long polyethylene tube, which was used for testing of the mole penetrators up to the depth of 5 m. The internal diameter of the tube is 0.74 m. Large diameter is necessary to minimize the effect of sand or lunar regolith simulant compression induced by the movement of the penetrator. The design of the test-bed system is presented in Fig. 4.

The main goal of the 2 m test-bed is to simulate different gravities for MOLE penetrator. The regular 5 m test-bed provides possibility of testing only for Earth's gravity. Changing inclination (β angle) enables us to test the mole in gravitational conditions of the Moon, Mars and other objects with mass smaller than Earth's; for example, for the Moon the angle of inclination equals 9.6°. However, the system cannot diminish the second order effect where the earth weight of the soil acts on the lateral wall of the KRET penetrator. The picture below shows the tube with the Moon regolith at inclination of β which can be expressed as a function of gravitation of the body g_{ob} in the following way:

$$\beta = 90^\circ - a\cos\left(\frac{g_{ob}}{g_{earth}}\right) \tag{1}$$

where g_{earth} is a gravitational constant at the Earth surface level (9.81 m/s^2) (Fig. 5).

The test-bed has ability to change inclination from 85 to 0°. In Fig. 5 these various angles are shown. The tube with regolith is mounted with two bearings on the structure made of steel frames. Hand chain block pulls up the tube. On the angle of 85°, the regolith is poured into the tube by a sand pump and then it is positioned to a given angle. The mole can dig itself up to 2.2 m.

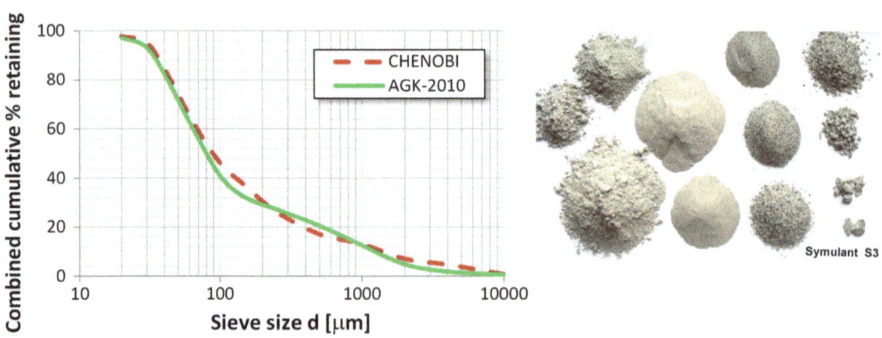

Fig. 3 Sieve analysis of the CHENOBI and AGK-2010 lunar analogue (*left*) and components of AGK-2010 (*right*)

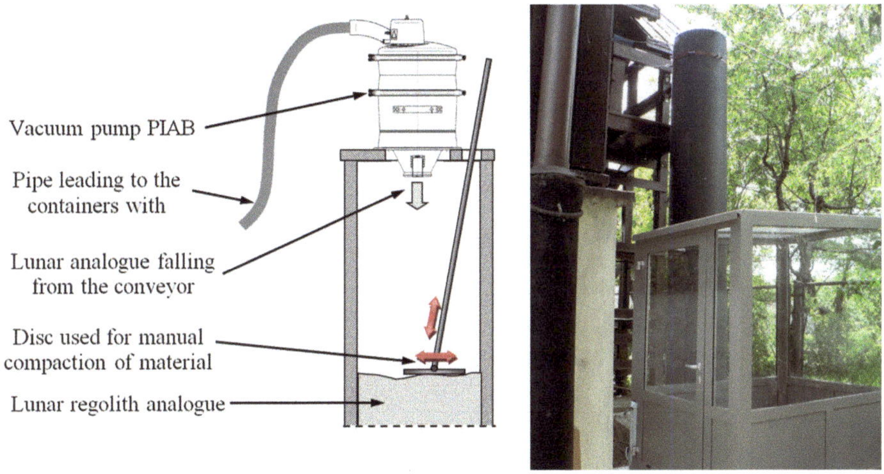

Fig. 4 The 5 m test-bed system. The schematic view together with the compaction system (*left*) and the real view of the test-bed system together with the control place (*right*)

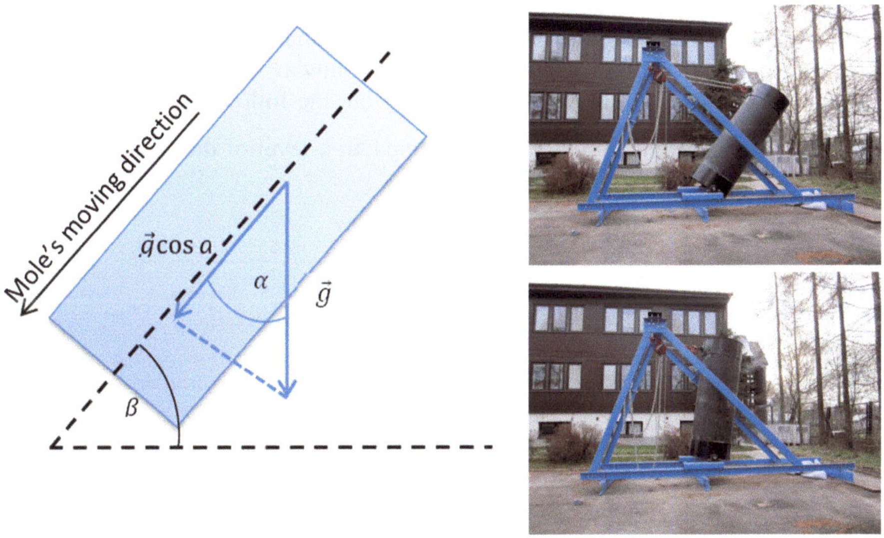

Fig. 5 Test-bed with various inclination angles. Schematic view on the *left*. The real view of the system in two different position on the *right*

4 Experimental Results

Successful tests of the mole penetrator in quartz sand and lunar regolith analogue AGK-2010 were performed in summer and autumn 2010. The main results are presented in Fig. 6 where the progress of the mole in quartz sand and lunar regolith

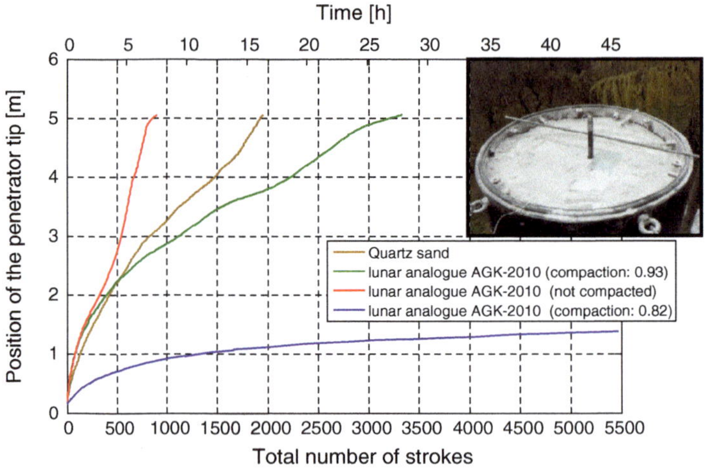

Fig. 6 Comparison of the results of mole penetrator KRET progress in 2 m test-bed with inclination of 45 and 9.6° with results from 5 m test-bed and results of mole progress with payload compartment

analogue with different compaction levels (ratio of material volume before compaction to material volume after compaction) is shown.

The results presented above allow pointing out the following conclusions:

- Mole progress depends highly on the compaction level of the lunar analogue.

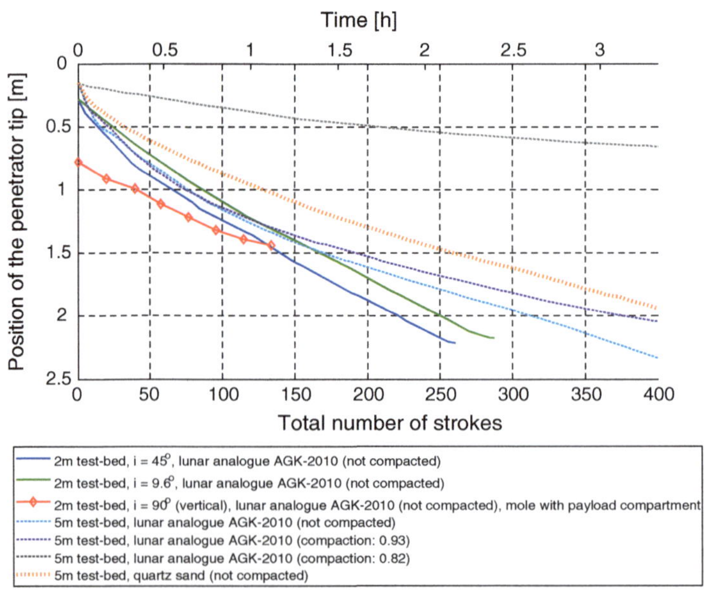

Fig. 7 Results of mole penetrator KRET progress in 5 m test-bed

Table 2 Summary of the mole KRET tests results

No	Material	Av. progress (mm/stroke)	Av. velocity (mm/h)	Final velocity (mm/h)
1	Quartz sand	2.59	310.4	319.8
2	AGK—2010 (not compacted)	5.72	665.9	528.2
3	AGK—2010 (0.93 compaction)	1.59	180.8	48.9
4	AGK—2010 (0.82 compaction)	0.26	30.2	6.8

- Mole progress decreases up to ~1.5 m depth and then remains constant, which means that the mechanical resistance of the cable is negligible.
- Stroke energy of the mole is an important parameter when working in highly compacted materials.
- Performed experiments confirm that the mole penetrator is able to work in subsurface layers of planetary bodies assuming their surface is made of regolith material.
- The tests results of the mole KRET in 2 m test-bed system with reduced gravity show that the gravity condition does not affect the overall performance of the mole KRET (see Fig. 7).

The summary of the results obtained during test campaigns are presented in the Table 2. It is worth to mention that at the end of the test the mole progress was not zero, which indicates that the mole KRET could penetrate the medium deeper.

5 Conclusions

The KRET penetrometer was design to work in regolith-like material which is typical for moons' and planets' sub-surface material in the Solar System. Typically it is used as measurement unit which allows determining the mechanical properties of the regolith. Furthermore, its features allow using it as a transportation device of different sensors (thermal sensors micro spectrometers). The experimental results presented in this chapter show the ability of the KRET penetrometer to drive itself up to 5 m deep into the lunar analogue. In case of the highly compacted material the mole slows down to couple of mm per hour which clearly means that the energy of the mole (assuming its fixed dimensions) should maximized. The testbed system with different inclination allows checking the mole behavior in reduced gravity. The obtained results are similar for all gravity conditions which is a good indicator of the expected penetrator behavior in the planetary environment. However, it is worth to mention that further investigations are needed to prove this feature. One of the possible solutions would be tests during parabolic flight.

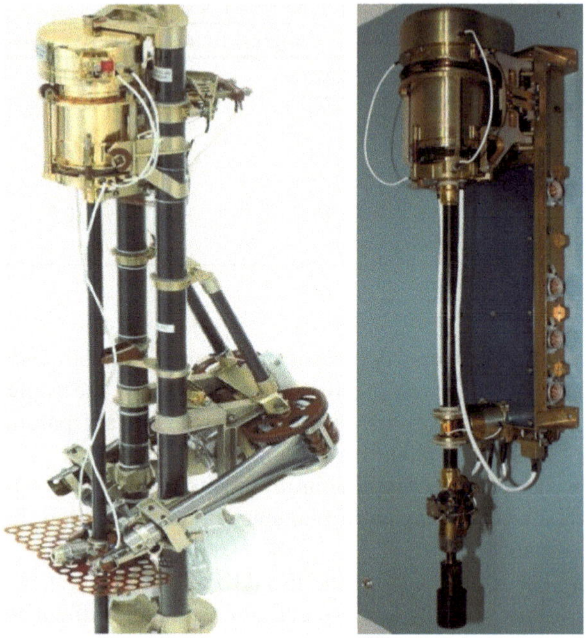

Fig. 8 MUPUS and CHOMIK penetrators—flight versions

During the long history of SRC involvement in penetrators development, several penetrators were successfully launched as a part of planetary missions (Fig. 8):

- MUPUS on the Philae lander (Rosetta mission) to 67P/Churyumov–Gerasimenko comet—landing time—2014—main tasks: measurements of thermal and mechanical properties of the comet subsurface layers.
- CHOMIK (Eng. hamster) on the Phobos Grunt mission—main tasks: sampling of the Phobos regolith and measurements of the thermal and mechanical properties. Unfortunately, the Phobos Grunt mission failed shortly after launch.

All penetrometers increase the knowledge of the subsurface layers in context of the future planetary bodies' exploration (for example lunar base development).

Acknowledgments The chapter was supported by ESA PECS project no. 98103.

References

Banaszkiewicz M, Seweryn K, Wawrzaszek R (2007) A sensor to perform in situ thermal conductivity determination of cometary and asteroid material. Adv Space Res 40(2):226–237

Coste P, Gelmi R, Magnani P, Poletto F, Schleifer A, Perrone A, Salonico A, Re E, Jollet D, Marson I, Corubolo P (2010) 'MOONBIT'—seismic while drilling (SWD) laboratory testing with Lunar regolith simulant. In: Proceedings of the 12th international conference on engineering, science, construction, and operations in challenging environments 'earth and space 2010', Honolulu, Hawaii

Dąbrowski B, Banaszkiewicz M (2006) Measurements of lunar and martian regolith thermal properties using subsurface robotic teams. In: Proceedings of the 9th ESA workshop on advanced space technologies and automation 'ASTRA 2006', ESTEC, Noordwijk

Grygorczuk J, Banaszkiewicz M, Seweryn K, Spohn T (2007) MUPUS insertion device for the Rosetta mission. J Telecommun Inf Technol 1(2007):50–53

Grygorczuk J, Seweryn K, Wawrzaszek R, Banaszkiewicz M (2008) Insertion of a mole penetrator—experimental results. In: Proceedings of the 39th lunar and planetary science conference, League City, Texas

Grygorczuk J, Seweryn K, Wawrzaszek R, Banaszkiewicz M (2009) Technological features in the new mole penetrator "KRET". In: Proceedings of the 13th European space mechanisms and tribology symposium, Vienna, Austria

Grygorczuk J, Banaszkiewicz M, Cichocki A, Ciesielska M, Dobrowolski M, Kędziora B, Krasowski J, Kuciński T, Marczewski M, Morawski M, Rickman H, Rybus T, Seweryn K, Skocki K, Spohn T, Szewczyk T, Wawrzaszek R, Wiśniewski Ł (2011a) Advanced penetrator and hammering sampling devices for planetary body exploration. In: Proceedings of the 11th ESA workshop on advanced space technologies for robotics and automation 'ASTRA 2011', ESTEC, Noordwijk, The Netherlands

Grygorczuk J, Dobrowolski M, Wisniewski L, Banaszkiewicz M, Ciesielska M, Kedziora B, Seweryn K, Wawrzaszek R, Wierzchon T, Trzaska M, Ossowski M (2011b) Advanced mechanisms and tribological tests of the hammering sampling device CHOMIK. In: Proceedings of the 14th European space mechanisms and tribology symposium (ESMATS), Constance, Germany

Seweryn K, Grygorczuk J, Wawrzaszek R, Rybus T, Wiśniewski Ł, Gonet A, Bednarz S, Rzyczniak M (2011) Lunar regolith analogue (in Polish: analog gruntu księżycowego). Patent application no. P397651

Spohn T, Seiferlin K, Hagermann A, Knollenberg J, Ball AJ, Banaszkiewicz M, Benkhoff J, Gadomski S, Grygorczuk J, Hlond M, Kargl G, Kührt E, Kömle N, Marczewski W, Zarnecki JC (2007) MUPUS—a thermal and mechanical properties probe for the Rosetta lander PHILAE. Space Sci Rev 128:339–362

Tulej M, Iakovleva M, Leya I, Wurz P (2011) A miniature mass analyser for in situ elemental analysis of planetary material–performance study. Anal Bioanal Chem 399:2185–2200

MIX
Papier aus verantwortungsvollen Quellen
Paper from responsible sources
FSC® C105338

If you have any concerns about our products,
you can contact us on
ProductSafety@springernature.com

In case Publisher is established outside the EU,
the EU authorized representative is:
Springer Nature Customer Service Center GmbH
Europaplatz 3, 69115 Heidelberg, Germany

Printed by Libri Plureos GmbH
in Hamburg, Germany